Communications in Computer and Information Science 2919

Series Editors

Rationale

The CCIS series is devoted to the publication of proceedings of computer science conferences. Its aim is to efficiently disseminate original research results in informatics in printed and electronic form. While the focus is on publication of peer-reviewed full papers presenting mature work, inclusion of reviewed short papers reporting on work in progress is welcome, too. Besides globally relevant meetings with internationally representative program committees guaranteeing a strict peer-reviewing and paper selection process, conferences run by societies or of high regional or national relevance are also considered for publication.

Topics

The topical scope of CCIS spans the entire spectrum of informatics ranging from foundational topics in the theory of computing to information and communications science and technology and a broad variety of interdisciplinary application fields.

Information for Volume Editors and Authors

Publication in CCIS is free of charge. No royalties are paid, however, we offer registered conference participants temporary free access to the online version of the conference proceedings on SpringerLink (http://link.springer.com) by means of an http referrer from the conference website and/or a number of complimentary printed copies, as specified in the official acceptance email of the event.

CCIS proceedings can be published in time for distribution at conferences or as post-proceedings, and delivered in the form of printed books and/or electronically as USBs and/or e-content licenses for accessing proceedings at SpringerLink. Furthermore, CCIS proceedings are included in the CCIS electronic book series hosted in the SpringerLink digital library at http://link.springer.com/bookseries/7899. Conferences publishing in CCIS are allowed to use our online conference service (Meteor) for managing the whole proceedings lifecycle (from submission and reviewing to preparing for publication) free of charge.

Publication process

The language of publication is exclusively English. Authors publishing in CCIS have to sign the Springer CCIS copyright transfer form, however, they are free to use their material published in CCIS for substantially changed, more elaborate subsequent publications elsewhere. For the preparation of the camera-ready papers/files, authors have to strictly adhere to the Springer CCIS Authors' Instructions and are strongly encouraged to use the CCIS LaTeX style files or templates.

Abstracting/Indexing

CCIS is abstracted/indexed in DBLP, Google Scholar, EI-Compendex, Mathematical Reviews, SCImago, Scopus. CCIS volumes are also submitted for the inclusion in ISI Proceedings.

How to start

To start the evaluation of your proposal for inclusion in the CCIS series, please send an e-mail to ccis@springer.com

Vladimir Jordan · Ilya Tarasov · Ella Shurina ·
Nikolay Filimonov · Vladimir Faerman
Editors

High-Performance Computing Systems and Technologies in Scientific Research, Automation of Control and Production

14th International Conference, HPCST 2024
Barnaul, Russia, May 17–18, 2024
Revised Selected Papers

Editors
Vladimir Jordan
Altai State University
Barnaul, Russia

Ilya Tarasov
MIREA – Russian Technological University
Moscow, Russia

Ella Shurina
Novosibirsk State Technical University
Novosibirsk, Russia

Nikolay Filimonov
Lomonosov Moscow State University
Moscow, Russia

Vladimir Faerman
Tomsk State University of Control Systems
and Radioelectronics
Tomsk, Russia

ISSN 1865-0929 ISSN 1865-0937 (electronic)
Communications in Computer and Information Science
ISBN 978-3-032-20324-3 ISBN 978-3-032-20325-0 (eBook)
https://doi.org/10.1007/978-3-032-20325-0

This Springer imprint is published by the registered company Springer Nature Switzerland AG
The registered company address is: Gewerbestrasse 11, 6330 Cham, Switzerland

Preface

The 14th International Conference on High-Performance Computing Systems and Technologies in Scientific Research, Automation of Control and Production (HPCST 2024) was held at Altai State University on May 17–18, 2024. Altai State University is located in the center of Barnaul, the capital of the Altai region in southwestern Siberia.

HPCST is an annual scientific event, held every year since 2011. It brings together specialists from various fields of modern computer and information science, as well as researchers working on applications in automation of control and production, mathematical modeling, and computer simulation of processes and phenomena in the natural sciences using parallel computing technologies.

The conference aims to present state-of-the-art approaches and methods for addressing contemporary scientific challenges. It also provides a platform for exchanging recent research results among scientists from universities and research institutions. All reported results represent valuable contributions to the field of applied information and computer science.

Conference sessions focused on the following topics:

- Architecture and design features of high-performance computing systems;
- Digital signal processors (DSP) and their applications;
- IP cores for field-programmable gate arrays (FPGA);
- Technologies for distributed computing using multiprocessors;
- Grid technologies, cloud computing, and related services;
- High-performance and multiscale predictive computer simulation;
- Computing in information security services;
- Control automation and mechatronics.

For the first time, the conference was held at an international level in 2017, attracting more than 160 researchers from Russia, China, Ukraine, Kazakhstan, Kyrgyzstan, Uzbekistan, Tajikistan, Vietnam, and Brazil. For many years, the average number of participants per year was around 60. At its peak in 2021, the conference attracted over 140 scholars, both virtually and in person.

Over the past few years, international participation has been affected. This year, the conference welcomed 83 scholars, with 81 reports accepted. These reports were selected by the Program Committee from 92 qualified submissions; 11 additional papers were declined during the routine editorial check before consideration. Among the participants, six reports came from Kyrgyzstan, Kazakhstan, and China, accounting for a total of seven international participants. The distribution of domestic participants remained largely unchanged, with the majority coming from the Siberian regions and Moscow.

The most significant and high-quality studies were carefully reviewed and included in this volume. Fewer than half of the reported papers were invited for these internationally published post-conference proceedings. Authors of 32 reports accepted the opportunity to contribute to this volume. Ultimately, the editorial board received 29 papers that had been revised to include internationally accessible references, extended to full-paper format, and prepared in accordance with the ethical standards of academic publishing. These manuscripts then underwent a second stage of peer review. Out of the 29 received papers, 6 were rejected by reviewers, primarily due to inconsistent discussion of results, low significance, or insufficient cooperation from the authors during the revision stage.

Following this review, 23 full papers presenting original studies in computing, mathematical simulation, control science, and information security were accepted for publication. Among them, 13 papers were accepted after minor revisions and 10 after major revisions. No papers were accepted without revision, and no guest papers were included in this volume.

During the extended period in which the volume remained in preparation, 4 papers were explicitly retracted by their authors or published elsewhere. Several authors requested revisions to provide updated results, reviews, and discussion. Accordingly, the content was updated prior to publication, with additional editorial checks and careful attention from the reviewers.

This volume contains 19 papers covering the following topics:

- Hardware for High-Performance Computing and Signal Processing;
- Information Technologies and Computer Simulation of Physical Phenomena;
- Computing Technologies in Data Analysis and Decision-Making;
- Information and Computing Technologies in Automation and Control Science;
- Computing Technologies in Information Security Applications.

To select the best papers presented at the conference, the procedure traditionally followed in the conference series was applied:

1. Session chairs prepared a shortlist of the most significant original reports that showed clear potential to be extended into full-paper format;
2. The editorial board, comprising the session chairs and the corresponding editor, compiled a list of up to 35 items;
3. Authors of the selected manuscripts were contacted and asked to extend and improve their papers before resubmitting them for review.

Every paper, without exception, was reviewed by at least three experts. In addition to a routine plagiarism check with Turnitin.com, an additional check was conducted using Antiplagiat.ru. This ensured that papers already published elsewhere in Russian were identified and excluded. Extended papers were allowed to show a certain level of similarity (up to 30%) to the corresponding conference materials previously published in Russian.

A single-blind review method was applied. The review criteria included:

1. Technical content;
2. Originality;
3. Clarity;
4. Significance;
5. Presentation style;
6. Ethics.

The organizing committee expresses sincere appreciation to the administration of Altai State University and the staff of the Institute of Digital Technologies, Electronics and Physics of Altai State University. The outstanding efforts of the technical staff made the conference possible despite travel restrictions.

The editors would also like to express their deep gratitude to the Springer editorial and production teams for providing the opportunity to publish the best papers as post-conference proceedings and for their excellent work on this volume.

January 2026

Vladimir Jordan
Ilya Tarasov
Ella Shurina
Nikolay Filimonov
Vladimir Faerman

Organization

General Chair

Vladimir Jordan	Altai State University, Russia

Program Committee Chairs and Section Charis

Ella Shurina	Novosibirsk State Technical University, Russia
Ilya Tarasov	Russian Technological University, Russia
Nikolay Filimonov	M.V. Lomonosov Moscow State University, Russia
Vladimir Faerman	Tomsk State University of Control Systems and Radioelectronics, Russia

Organizing Committee

Vasiliy Belozerskih	Altai State University, Russia
Alexander Kalachev	Altai State University, Russia
Vladimir Pashnev	Altai State University, Russia
Viktor Sedalischev	Altai State University, Russia
Yana Sergeeva	Altai State University, Russia
Igor Shmakov	Altai State University, Russia
Petr Ulanov	Altai State University, Russia

Program Committee

Viktor Abanin	Biysk Technological Institute, Russia
Valeriy Avramchuk	Tomsk State University of Control Systems and Radioelectronics, Russia
Sergey Beznosyuk	Altai State University, Russia
Alexander Filimonov	MIREA – Russian Technological University, Russia
Pavel Gulyaev	Yugra State University, Russia
Ishembek Kadyrov	Kyrgyz National Agrarian University, Kyrgyzstan
Alexander Kalachev	Altai State University, Russia

Vladimir Khmelev	Polzunov Altai State Technical University, Russia
Shavkat Fazilov	Research Institute for the Development of Digital Technologies and Artificial Intelligence, Uzbekistan
Vladimir Kosarev	Khristianovich Institute of Theoretical and Applied Mechanics SB RAS, Russia
Nomaz Mirzaev	University of Information Technologies n.a. al-Khorezmi, Uzbekistan
Elena Kruchkova	Polzunov Altai State Technical University, Russia
Aleksey Nikitin	Altai State University, Russia
Viktor Polyakov	Altai State University, Russia
Aleksey Yakunin	Polzunov Altai State Technical University, Russia
Dmitriy Potekhin	MIREA – Russian Technological University, Russia
Sergey Pronin	Polzunov Altai State Technical University, Russia
Anatoliy Gulay	Belarus National Technical University, Belarus
Gambar Guluev	Institute of Control Systems of Azerbaijan National Academy of Sciences, Azerbaijan
Viktor Sedalischev	Altai State University, Russia
Vitaliy Titov	Southwest State University, Russia
Pedro Filipe do Prado	Federal University of Espirito Santo, Brazil
Bruno Pissinato	Methodist University of Piracicaba, Brazil

External Reviewers

Alexey Saveliev	Tomsk Polytechnic University, Russia
Alexey Tsavnin	Tomsk Polytechnic University, Russia
Andrey Malchukov	Tomsk State University of Control Systems and Radioelectronics, Russia
Anton Kiselev	Siberian State University of Telecommunications and Informatics, Russia
Anton Konev	Tomsk State University of Control Systems and Radioelectronics, Russia
Anton Yants	Perm National Research Polytechnic University, Russia
Bahyt Absadykov	Kazakh-British Technical University, Kazakhstan
Bibigul Koshoeva	Razzakov Kyrgyz State Technical University, Kyrgyzstan
Daria Novokhrestova	Tomsk State University of Control Systems and Radioelectronics, Russia
Elena Luneva	Tomsk State University of Control Systems and Radioelectronics, Russia

Contents

Hardware for High-Performance Computing and Signal Processing

Characteristics of Processor Components Implemented as a Part of FPGA

Ilya Tarasov(✉)

MIREA –Russian Technological University, Vernadsky Avenue 78, 119454 Moscow, Russia
tarasov_i@mirea.ru

Abstract. The article analyzes the development features of processor components implemented as part of a programmable logic integrated circuit (FPGA). The FPGA architecture features determine the use of optimal design techniques that use recommended operating modes of logic cells and specialized hardware components. The article considers the characteristics of typical processor components synthesized on the basis of logic cells and memory components included in FPGA chips. It is determined that the size of the register file implemented on the basis of FPGA cells is significantly larger than the size of the arithmetic logic unit supporting the main list of operations and implemented on the basis of the same cells. Therefore, a promising way to develop processor architectures for FPGA is to complicate the arithmetic logic unit by cascading its submodules. This approach should combine the study of typical command sequences that can be generated by compilers for the subject areas of the processor. In addition, the complexity of the arithmetic logic device is limited by the features of the architecture of the FPGA logic cells.

Keywords: FPGA · Architecture · Register File · Arithmetic Logic Unit · Processor

1 Introduction

In practice, the requirements for processor modules implemented as part of FPGA-based projects should differ from the requirements for processors that are the central computing devices in the system. FPGAs are characterized by high performance of subsystems based on hardware components, such as DSP modules, transceivers, and computing nodes with a parallel architecture synthesized on the basis of logical cells. In this case, processor cores are used mainly to control the operation of specialized components, ensuring a reduction in development time due to the replacement of the operations of synthesizing a modified circuit with the correction of the software of the built-in processor cores. In [1], a system based on more than 330 FPGAs with FPGA architecture and 3800 FPGAs with CPLD architecture is mentioned, in which more than 300 32-bit Microblaze processor cores and several thousand simpler 8-bit Picoblaze processors are used for individual control tasks. In this case, the simplicity of project support and refinement of operating algorithms by quickly replacing the software

V. Jordan et al. (Eds.): HPCST 2024, CCIS 2919, pp. 3–14, 2026.
https://doi.org/10.1007/978-3-032-20325-0_1

of embedded processors comes to the fore. Achieving peak performance is not the most important task for such cores, since the system performance is determined by the characteristics of specialized computing modules. Reducing the power consumption is also irrelevant, since it is largely determined by the characteristics of the FPGA used. Thus, for processors synthesized on the basis of FPGA logic cells (the so-called soft processors), attention should be paid to the simplicity of integration into the project, predictable performance characteristics (with moderate values allowed for performance), and, if possible, compact program code [2].

The processor in the FPGA can be used for digital signal processing systems implemented on the basis of hardware components. In this scheme, the processor is not the main node for processing the data flow. Instead, the main performance is provided by nodes assembled on the basis of DSP48 components, and the processor solves auxiliary problems - setup, calibration and monitoring of operation, as well as organizing the interface with the operator and external chips. Therefore, the control processor of the DSP system does not require high computing performance, and instead the main emphasis can be placed on closer integration with digital signal processing modules. In embedded systems, it is also possible to indicate the areas of preferred use of the soft processor [3]. In applications with a large number of ongoing processes of a heterogeneous nature, it is often necessary to organize the work of the microcontroller around interrupt handlers. Linear execution of the main algorithm is combined with the processing of interrupt requests from various external sensors and peripheral devices of the microcontroller itself. For the programmer, this behavior may ultimately prove inconvenient to maintain, and the simultaneous arrival of several interrupt requests may lead to an uncertain or even emergency situation.

With this approach, auxiliary tasks that require intensive calculations are completely transferred to the coprocessors. The central processor, as in the case of digital signal processing systems, performs mainly the functions of control, monitoring, and organizing external interfaces.

If the processor plays a supporting role, the requirements for it differ from those for an application processor. Computing performance may be average, because in FPGAs, the bulk of operations are performed by hardware components and specialized IP cores. FPGA power efficiency is primarily determined by the family and the manufacturing process on which it is based. FPGA power management is limited, and many power optimization techniques typical of VLSI are unavailable when designing with FPGAs. Furthermore, reducing the power consumption of a limited portion of the FPGA will not impact other components, which typically contribute the most to the design's power consumption. Therefore, special measures to reduce power consumption for control software processors are impractical.

On the other hand, control tasks are often IO-oriented. Therefore, the control processor must be tightly integrated with peripheral components and capable of quickly performing operations to validate and transform data from peripheral controllers. Operations such as bit masking, combining and permuting bit fields, and calculating complex conditions generated by multiple peripheral devices may be required. Such operations generate a large set of combinations that cannot be supported equally in a general-purpose processor but can be implemented as specialized instructions in a software processor.

Therefore, the challenge is to develop a software processor that is tightly integrated with peripheral devices and capable of executing specialized instructions designed to support a strictly defined set of hardware controllers and operations with them. Therefore, the characteristics of the main components of the soft-processor will differ from those typical for general-purpose processors.

2 Components of Simple Processor Cores for FPGA

Figure 1 shows the architecture of a processor core based on a two-stage pipeline. It is a simple example of a working processor that contains the main components that ensure its operation. In accordance with the Harvard architecture, program memory and data memory are separated, and a register file and an arithmetic logic unit (ALU) are implemented for the data path. Simple processor architectures are discussed in detail in a number of literary sources [4, 5]. The components shown can be perceived during development as somewhat equal in terms of their influence on the final characteristics of the processor core in target tasks. At the same time, their characteristics are also improved independently from the point of view of module development, but both the register file and the ALU have different mechanisms for influencing the processor performance, and from the point of view of hardware implementation, on the share of FPGA resources occupied by this processor. In this regard, it is important to evaluate the influence of the parameters of the main components of the processor specified during development on the volume of resources occupied by them, with the aim of further joint optimization of individual modules [6].

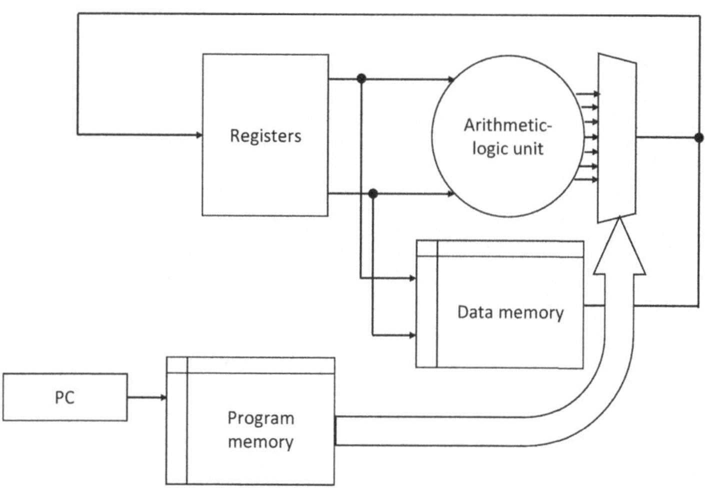

Fig. 1. Simple processor core architecture.

Further in the article, examples of the implementation of individual components are considered in order to determine their logical capacity, as well as the dependence of the occupied FPGA resources on the main parameters that determine the characteristics of the processor core. When assembling processor components, it is advisable to balance the resources used to achieve the greatest efficiency of the design as a whole, rather than its individual components.

3 Implementation of the Register File in FPGA

For a general-purpose register file with symmetric register access, an obvious approach to its RTL description can be taken using an array of signals. For the example shown, a non-parameterized description with 32 registers is used, with write signals implemented separately for the low and high 16-bit words.

```
module regfile(
input clk,
input reset,
input [4:0] aw,
input [31:0] d,
input [1:0] we,
input [4:0] opa,
input [4:0] opb,
output [31:0] qa,
output [31:0] qb);

reg [31:0] regs[31:0];

integer i;

always @(posedge clk)
begin
  if (we[0]) regs[aw][15:0] <= d[15:0];
  if (we[1]) regs[aw][31:16] <= d[31:16];
end

assign qa = regs[opa];
assign qb = regs[opb];

endmodule
```

A comparative analysis of the characteristics of register files synthesized for several FPGA families revealed significant differences in the results. The primary driver of these differences is the availability of dedicated logic cell configuration modes, which can be used in distributed memory mode on Xilinx FPGAs. This mode allows for the implementation of a register file with a 16x1 or 32x1 organization in a single logic generator (LUT) enabled in distributed memory mode. Therefore, simple register files offer a significant advantage in terms of compact implementation for Xilinx FPGAs.

The results of register file synthesis with different configuration options for FPGA of several families are given in Table 1.

Table 1. Comparative analysis of the results of synthesis of register files with 2, 3 and 4 ports for reading

Read ports	2	3	4
Xilinx Artix-7, reset is used			
FFs	1024	1024	1024
LUT	1743	2031	2319
Xilinx Artix-7, reset is not used (distributed memory mode is allowed)			
FFs	0	0	0
LUT (LUTRAM mode)	128	192	256
Intel Cyclone V			
FFs	1024	1024	1024
LUT	608	1091	1444

An analysis of the table shows that the register file implementation method significantly impacts the required FPGA resources. Therefore, in FPGAs supporting distributed memory mode (distributed RAM in AMD, shadow RAM in Gowin), implementing a large register file will not lead to an unjustified increase in hardware resources.

A separate type of register file is the stack-based register organization. It has no hardware storage resources, but with a zero-operand instruction set, no operands need to be specified in the instruction, because the operands and result are determined by the current position of the top of the stack.

4 Pipeline Arithmetic-Logic Unit in FPGA CAD

The arithmetic logic unit (ALU) is the processor's main computing unit, performing data transformations. An ALU performs operations on input operands (obtained from the register file or data memory) based on the instruction code being executed. Accordingly, the ALU implementation is quite simple.

```
library IEEE;
use IEEE.STD_LOGIC_1164.ALL;
use IEEE.NUMERIC_STD.ALL;

entity alu is
    Port ( opa : in STD_LOGIC_VECTOR (31 downto 0);
           opb : in STD_LOGIC_VECTOR (31 downto 0);
           din : in STD_LOGIC_VECTOR (31 downto 0);
           cmd : in STD_LOGIC_VECTOR (31 downto 0);
           result : out STD_LOGIC_VECTOR (31 downto 0));
  end alu;

  architecture Behavioral of alu is

  signal add, sub, mult : integer;

  begin
  add         <=          to_integer(unsigned(opa))          +
 to_integer(unsigned(opb));
  sub         <=          to_integer(unsigned(opa))          -
 to_integer(unsigned(opb));
  mult         <=          to_integer(unsigned(opa))          *
 to_integer(unsigned(opb));

  with cmd(27 downto 24) select
  result <= opa when "0000",
      std_logic_vector(to_unsigned(add, 32)) when "0001",
      std_logic_vector(to_unsigned(sub, 32)) when "0010",
      opa and opb when "0011",
      opa or  opb when "0100",
      opa xor opb when "0101",
      opa(30 downto 0) & '0' when "0111",
      '0' & opa(31 downto 1) when "1000",
      opa(31) & opa(31 downto 1) when "1001",
      not(opa) when "1010",
      din when "1011",
      opa(31 downto 16) & cmd(15 downto 0) when "1100",
      cmd(15 downto 0) & opa(31 downto 16) when "1101",
      std_logic_vector(to_unsigned(mult, 32)) when "1111",
      (others => '0') when others;

  end Behavioral;
```

The synthesis result is shown in Fig. 2. It can be seen that the ALU is a simple multiplexer controlled by the instruction code (or at least part of this code). In this example, the type of operation being executed is specified in bits (27..24) of the instruction code.

Signal propagation delay depends on the operation category performed in the ALU. For FPGAs, the following categories can be distinguished, implemented by different components of the FPGA configurable resource matrix.

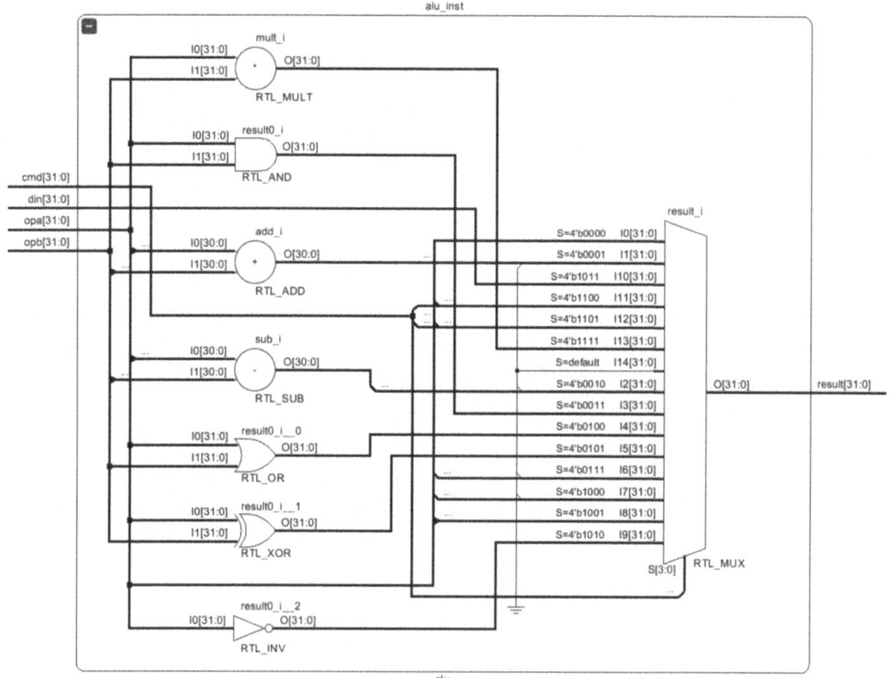

Fig. 2. Arithmetic-logic unit (elaborated view)

Bitwise operations such as AND, OR, XOR, as well as bit shift and combine operations, are implemented using LUTs and logic cell multiplexers. These operations have a fixed delay defined by the FPGA family, independent of the operation type. The increased delay is due to the addition of additional cells in the critical signal propagation path, which is determined not by the type of operations performed, but by their number.

Addition and subtraction operations are implemented using fast carry chains, which are included in logic cells as separate, independent resources and do not require the use of a LUT. These resources are more efficient than LUTs and should not be prevented from being detected by the synthesizer, as optimization techniques based on the transformation of combinational expressions will lead to the synthesis of a LUT, so the effect of the optimization will only be apparent.

The multiplication operation is implemented using hardware blocks found in many FPGA families. It is also an operation that is detected by the synthesizer to use a dedicated hardware block. The location of the multiplication blocks in the FPGA is fixed, so adding a multiplication operation to the ALU creates additional trace lines from the multiplication block output to the ALU output multiplexer. Depending on the FPGA family, using multiplication can reduce the overall processor clock rate because the path through the multiplication block is the critical path [8].

Pipelining the multiplier path is a useful practice to increase the processor clock rate, but it requires hardware data dependency resolution or a special compiler condition to delay the use of the multiplication result.

According to [9], the importance of preliminary research of the subject area can be indicated in order to avoid redundant ALU composition.

An experiment with data path pipelining was conducted by eliminating the multiplication instruction from the ALU. While hardware multipliers in FPGAs enable high clock rates, it's important to consider that such a unit is located separately, requiring signal routing from its output to the ALU multiplexer. Target platform: Artix-7 FPGA.

Two stages without multiplication: 9.534 ns (105 MHz).

With registers in the Decode stage without multiplication: 6.043 ns (165 MHz).

After data advancement without multiplication: 7.076 ns (141 MHz).

5-stage pipeline without multiplication: 6.567 ns (152 MHz).

Two factors can be seen influencing clock rate variation. The first and most expected is the number of pipeline stages. The significant increase in performance from two to three stages is due to the splitting of the critical path from the register file through the ALU to the register file write input. However, the transition to five pipeline stages did not result in further performance gains. This is because the critical path became the circuit within the ALU. Eliminating the multiplication unit resulted in an increase in clock frequency, even within a single pipeline architecture.

This impact of the multiplication operation is not due to the low clock frequency of the hardware multiplication unit, but to the need to route signals to its inputs and from the output to the ALU multiplexer. The frequency at which the multiplication unit can operate, as specified in the technical documentation, is 375 or 450 MHz for Artix-7 FPGAs of various speed grades. This frequency is achieved if the unit's input and output signals are written to embedded registers. In this case, the critical path can be the circuit transferring input operands from the sources or the result to the ALU output. Therefore, further pipelining should be considered within the arithmetic logic unit, achieving not only a reduction in the depth of series-connected logic cells, but also a compact arrangement of connected FPGA resources.

If the multiplication operation is rarely used in control algorithms, it can be replaced with a sequential multiplication, for which support for the "shift and add" instruction is sufficient. This eliminates the need for a hardware multiplier, but the impact on the performance of the soft processor is ambiguous. Eliminating a dedicated hardware component replaces it with FPGA logic cells, which increases the complexity of the ALU. Furthermore, the trace no longer captures the hardware multiplier with its fixed location on the chip, which may be suboptimal relative to the ALU layout. The implementation of the multiplication operation should be validated experimentally by comparing the overall processor characteristics, testing not only the clock frequency but also the overall performance when executing domain-specific programs. If the multiplication operation is rarely used, the effect of reducing the clock frequency will be predominantly negative.

The paper [11] discusses a unified description of a computing node using four parameters (I, O, D, S), where:

- is the number of instructions executed per clock cycle;
- is the number of operations defined by the instruction;
- D is the number of operands (operand pairs) associated with the operations;
- S is the degree of pipelining.

For SIMD architectures, D > 1, and for MIMD (i.e., Multiple Instruction, Multiple Data) architectures, I > 1 and D > 1. For massive parallelism, we can additionally consider datapath parallelism at the level of an individual computing node, which allows us to apply an architectural description of the node as a quadruple (R, <O>, <D>, <S>), where:

- R is the number of independently computed results;
- <O> is the vector of the number of operations performed by the instruction for each datapath;
- <D> is the vector of the number of operands (operand pairs) for each datapath;
- <S> is the pipelining vector for each datapath.

This approach involves replacing the parameters O, D, and S with vectors describing the characteristics of the corresponding datapaths, which form R outputs.

The FPGA's capabilities for reconfigurable and parallel design open up vast opportunities for creating original circuits supporting arithmetic, logical, and special operations, as well as their wide combination. The study of register file characteristics, described earlier in the article, opens up the possibility of creating ALUs with 3, 4, and even more inputs, which, however, require adequate support from the development software. In some cases, when creating control software processors in FPGAs, operations can be specified in assembly language if they represent critical instruction sets that significantly improve the software's performance.

The following variants of a single data path can be considered as examples. Figure 3 shows a chain of computing units using 2 pairs of operands and 3 operations (O = 3, D = 2). If the data path using pipelining (S = 1), latency is generated in the data path, which must be taken into account in the instrumentation software or in the hardware formation of wait cycles in the processor. If we consider the shown examples as individual outputs of the combined ALU of the processor, its architectural description will have the form: (2, <3, 3>, <2, 2>, <0, 1>). This description does not consider possible variations in the composition of the computing nodes shown as ALU1, ALU2, ALU3, although their functionality may differ both within a single path and for individual paths.

Cascading the ALU can introduce additional delay, which has a greater negative impact on processor performance than the additional FPGA resources, which, when adding ALU submodules, are small compared to the overall processor size. Therefore, when improving the ALU, attention should be paid to parallel, rather than cascading, instruction execution capabilities.

The approach of parallel instruction execution on different pairs of operands corresponds to the VLIW (Very Long Instruction Word) architecture, where the instruction size is determined by the number of independent actions and operands used in the instruction. The VLIW approach can be effectively used in FPGAs because synthesizing memory blocks with wide data words presents no significant difficulty. However, program size growth is a more significant issue due to the limited memory capacity of FPGAs.

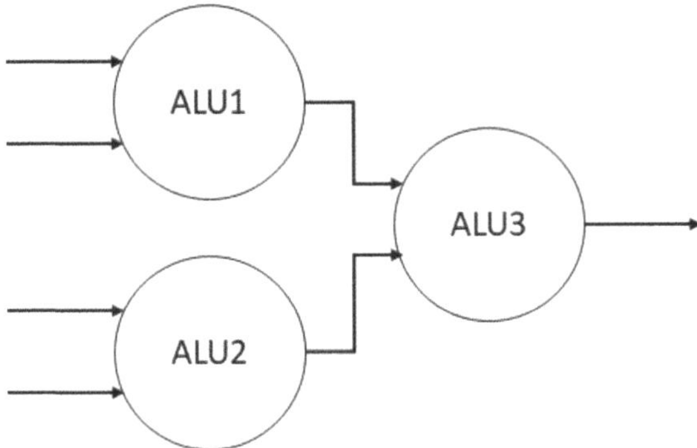

Fig. 3. Cascaded arithmetic-logic unit

For a processor with a VLIW architecture, the data paths used may not be identical. For example, one or more paths may be used to organize memory operations, generating an address and data for each of the memory access interfaces used. For example, the following program fragment contains operations that can be executed in parallel:

```
for (i = 0; i < 10; i++) x[i] = 0;

Pseudocode:
i = 0;
loop: x[i] = 0;
i = i + 1;
cmp i, 10
jless loop

Pseudocode version 2:
i = 0;
loop: x[i] = 0; i = i + 1; cmpjless loop
```

The actions shown in listing can be performed in an ALU containing three computational nodes implementing the corresponding operations performed in the loop. Figure 4 shows a data path that allows the loop body to be executed in a single cycle. The branch operation is not shown, as it pertains to the processor's control flow, not the data path.

The example considered represents a particular solution to the problem, since the resulting ALU structure is essentially designed to accelerate the single example considered. Repeating this approach will result in ALU structures that may differ from the one shown in Fig. 3. Therefore, it is necessary to consider an ALU design flow based on a formalized design methodology and suitable for creating ALUs specialized for a given set of domain area.

A promising direction for ALU development is accelerating access to peripheral components through parallel calculation of addresses, data, and read/write conditions. These operations, implemented in separate ALU submodules, can significantly reduce

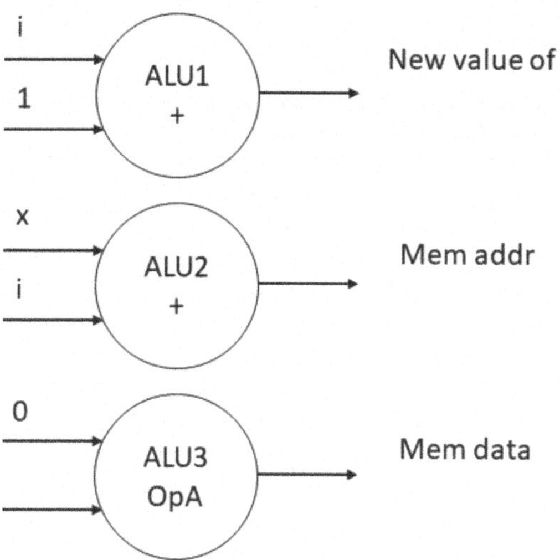

Fig. 4. VLIW-style arithmetic-logic unit

the number of processor cycles required to perform typical data exchange operations with peripheral controllers.

The importance of domain-specific programming languages, which can identify the instruction sets representable in the ALU for a VLIW processor, is growing. FPGAs' parallel processing capabilities open up new possibilities for direct control of processor components, which are impractical to implement in general-purpose processors due to the excessive number of possible instruction sets. This approach can be used in specialized software processors designed to control a predetermined set of peripheral controllers, where typical operations can be defined in advance.

5 Conclusions

The article demonstrates the dependence of the characteristics of computing system components on the FPGA family used. Therefore, the processor architecture and the pipelining depth of the arithmetic logic unit must be aligned with the characteristics of the hardware platform used to implement the processor. The importance of tight integration of the soft processor with other system components and the prospects of using the VLIW architecture to support parallel execution of commands with direct control of ALU submodules can be noted.

Acknowledgements. This work is supported by the Ministry of Science and Education of RF (Project No. FSFZ-2022-0004).

References

1. Hennessy, J.L., Patterson, D.A.: 2017 Computer architecture. 6th Edition. A Quantitative Approach (The Morgan Kaufmann Series in Computer Architecture and Design) (2017)
2. Harris, S., Harris, D.: Digital design and computer architecture: ARM edition, 1st edn. Morgan Kaufmann Publishers Inc., San Francisco, CA, USA (2015)
3. Currie, E.: Microcontroller subsystems. In: Mixed-Signal Embedded Systems Design, pp. 35–97. Springer, Cham (2021). https://doi.org/10.1007/978-3-030-70312-7_2
4. Tarasov, I.E.: Architecture of a Specialized VLSI for Serial Digital Signal Processing. In: Proceedings of the 3rd International Conference on Control Systems, Mathematical Modeling, Automation and Energy Efficiency, SUMMA 2021, vol. 3, pp. 457–460 (2021)
5. Tarasov, I.E., Potekhin, D.S.: VLSI architecture with a configurable data processing path based on serial distributed arithmetic. J. Phys: Conf. Ser. **1615**, 012001 (2020)
6. Balfour, J., Dally, W.J., Black-Schaffer, D., Parikh, V., Park, J.S.: An energy-efficient processor architecture for embedded systems. IEEE Comput. Archit. Lett. **7**(1), 29–32 (2008)
7. Jhamb, M., Lohani, H.: Design, implementation and performance comparison of multiplier topologies in power-delay space. Eng. Sci. Technol. Int. J. **19**(1), 353–363 (2015). https://doi.org/10.1016/j.jestch.2015.08.006
8. Kanduri, A., Rahmani, A.M., Liljeberg, P., Hemani, A., Jantsch, A., Tenhunen, H.: A Perspective on dark silicon. In: Rahmani, A., Liljeberg, P., Hemani, A., Jantsch, A., Tenhunen, H. (eds.) The Dark Side of Silicon. Springer, Cham (2017). https://doi.org/10.1007/978-3-319-31596-6_1
9. Hennessy, J.L., Patterson, D.A.: A new golden age for computer architecture: domain-specific hardware/software co-design, enhanced security, open instruction sets, and agile chip development. In: Proceedings of 2018 ACM/IEEE 45th Annual International Symposium on Computer Architecture (ISCA), pp. 27–29. Los Angeles, CA, USA (2018). https://doi.org/10.1109/ISCA.2018.00011
10. Varma, R.A.C., Subbarao, M.V., Varma, D.R., Raju, G.R.L.V.N.S.: High-Throughput VLSI architectures for VLSI signal processing. Lect. Notes Electric. Eng. **655**, 349–358 (2021). https://doi.org/10.1007/978-981-15-3828-5_37
11. Sima, D., Fountain, T., Kacsuk, P.: Advanced computer architectures - a design space approach. Addison-Wesley (1997)

Selecting Optimal Signal Transformation Block for Real-Time Speech Enhancement Using Deep Neural Network

Andrey Lependin(✉), Valentin Karev, and Pavel Zubkov

AltSU – Altai State University, Lenin Ave. 61, 656049 Barnaul, Russia
lependin@phys.asu.ru

Abstract. In this paper the comparison of several types of signal transform layers in speech enhancement using deep neural networks was done. The main requirements to be considered when choosing the type of transform were the quality of the enhanced signal, the size of the neural network model and the possibility of real-time operation. It was proposed to apply recurrent transformations of LSTM type and two types of state space transformations S4 and Mamba. The neural network consisted of two main blocks of transformations. The first one processed the whole spectrum of the signal and the second one worked with separate frequency bands. Data from the DNS Challenge set were used as training data. They were used to generate examples of distorted signals with signal-to-noise ratio in the range of 0–20 dB. The trained models were tested on a test subsample of the same set, as well as on synthesized samples with a given value of the signal-to-noise ratio in the range from −5 to 20 dB, in order to assess the generalization ability of the models. The results of numerical experiments showed that the LSTM-based and Mamba-based models performed the best. The latter had a significantly smaller number of trained parameters, which allowed us to consider it as the most suitable for improving the quality of speech signals.

Keywords: Speech Technologies · Speech Enhancement · Noise Suppression · Noise Masking · Signal Processing · Deep Learning · State Space Model Layer · Recurrent Layer

1 Introduction

The proliferation of modern information systems using multimedia transmission has led to increased demands on the quality of the voice signal delivered to end users. The audio stream must be transmitted without delay and with minimal distortion. The speaker's voice must remain realistic, with natural timbre and intonation. At the same time, registration of the transmitted signal can take place in non-neutral conditions (at home, in the street, in a crowded office, etc.). In this case, there are a large number of noise sources around the user, which have a significantly non-stationary character. The room itself can cause sound reflections from free surfaces, distorting the speaker's voice. Classical noise reduction methods, based on the assumption of quasi-stability of noise

V. Jordan et al. (Eds.): HPCST 2024, CCIS 2919, pp. 15–27, 2026.
https://doi.org/10.1007/978-3-032-20325-0_2

sources, cannot cope with such a combination of quality requirements and complexity of distortions introduced into the signal. Therefore, modern developers of mass information and videoconferencing systems are extremely interested in the development of new, more advanced speech enhancement methods. This is indirectly confirmed by the wealth of work on noise or reverberation reduction and general improvement of audio signals that has been carried out in recent years, both on behalf of companies (Microsoft's DNS Challenge [1]) and within the framework of major international conferences (ICASSP 2023 Deep Noise Suppression Challenge [2]).

The widespread use of deep neural network learning has greatly accelerated speech enhancement methods. Neural networks are highly complex and have the ability to automatically identify significant patterns in the signal. This allows not only to reduce the impact of distortion in the processed speech signal, but also to restore individual parts of the signal to achieve the most realistic sounding result. Not all neural network architectures are equally suited to the task of speech enhancement. The size of the network should be small (no more than a few million parameters) and the speed of inference should be high (with a delay of no more than 30–40 ms). The requirement for high quality and real-time operation most often leads to the use of recurrent layers for signal processing [3, 4], possibly complemented by self-attention layers [5]. The use of state space model (SSM) transformations is promising [6, 7] and has shown good results in speech processing tasks [8].

In this paper, we compared the performance of several similar neural network models for speech enhancement, which differed only in the signal processing blocks used. The latter were recurrent transformation and two types of SSM layers (S4 [6] and Mamba [7]). The quality of the work was evaluated over a relatively wide range of signal-to-noise ratios of the distorted signals, using a wide range of metrics for evaluating the quality of speech enhancement.

2 Deep Neural Network Model for Speech Enhancement

2.1 Speech Enhancement with Complex Ideal Ratio Masks

There are two main types of distortion introduced into the recorded speech signal. The first type is associated with the mixing of various types of background noise into the audio signal [9]. Noise is any unwanted sound signal in the vicinity of a recording microphone, including various natural sounds, machine sounds, animal sounds and other people's voices. The second kind of distortion is multiplicative. The amount of distortion depends on the amplitude of the original signal and varies to different degrees at different frequencies. It manifests itself as reverberation or echo in the room when a speech signal is recorded [3]. Correcting multiplicative distortion is a more difficult task than removing additive noise because the former is associated with the original signal and can cause significant distortion of the restored audio signal. The introduction of both types of distortion can be represented as the following transformation of a clean signal s(t):

$$x(t) = y(t) + n(t) = s(t) * h(t) + n(t), \tag{1}$$

where x(t) is a distorted signal, y(t) is a reverberated signal; n(t) is additive noise, h(t) is the impulse response of the room in which the signal is recorded, * is the signal convolution operation. When decomposed into a time-frequency representation, Eq. (1) is transformed into the following form:

$$X(t, f) = Y(t, f) + N(t, f) = S(t, f) \cdot H(t, f) + N(t, f), \tag{2}$$

where the signals from (1) are replaced by the corresponding complex amplitudes of the Fourier transform for each discrete frequency f and an arbitrary signal frame with the number t, the convolution operation is replaced by a product.

The most common method of speech signal restoration is the use of masks for the ratio of complex amplitudes of spectral decompositions [10] of enhanced and distorted signals M(t, f). The neural network model predicts this mask for a given frame based on Fourier decompositions of previous signal frames and previous mask values. The operation of calculating the spectrum of the restored signal is written as:

$$S_e(t, f) = M(t, f) \odot X(t, f), \tag{3}$$

where $\odot$ is the element-wise product of the complex amplitudes of the corresponding frequency components of the mask and the signal. The resulting time-frequency representation $S_e(t, f)$ undergoes an inverse Fourier transform.

It should be noted that the effective removal of an additive noise requires that the amplitude mask values should have similar characteristics to the enhancement signal. Due to the fact that the natural scale for speech is the logarithmic (decibel scale), then the mask values must be distributed according to a significantly nonlinear law. Because of this, neural network models usually predict not the mask itself, but its so-called compressed representation [10] of the form:

$$\mathrm{Cmpr}(M(t, f))_{r,i} = K\big(1 - \exp(-CM(t, f)_{r,i})\big)/\big(1 + \exp(-CM(t, f)_{r,i})\big). \tag{4}$$

This transformation is performed separately for the real (r) and imaginary (i) components of the mask, which is emphasized by the use of subscripts. The compression parameters according to [10] took the values K = 10 and C = 0.1.

2.2 Neural Network Model Architecture

The structure of the neural network model architecture used in the work is shown in Fig. 1. It repeats the architecture used in previous work [11, 12] for real-time speech enhancement. The main difference is that the coding components of the network can be replaced by other transformations.

The processed audio signal is divided into frames of equal duration, following with a time step Δt. Each of the frames is pre-filtered, a windowing function is applied to it and the Fourier transform X(t, f) is calculated, where $f = 0,\ldots,F-1$ is the index of the discrete frequency (frequency band number) of the signal. The mask is calculated with a delay of τ frames, the processed signal is delayed by $\tau\Delta t$ behind the input distorted one.

The normalization of the real amplitude vectors in Fig. 1 is performed by dividing by the mean of the corresponding values of the previous vectors estimated by the moving

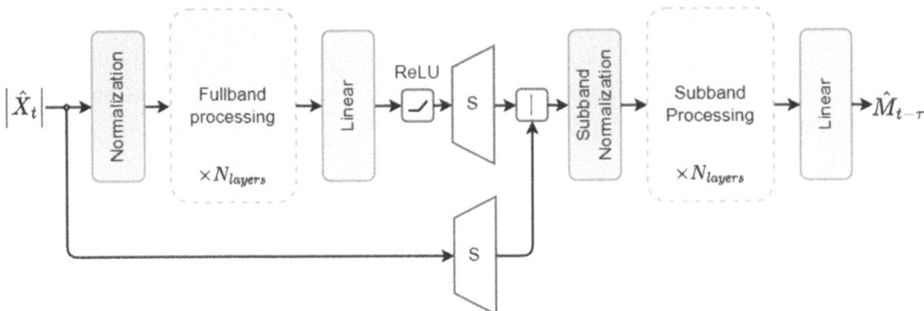

Fig. 1. The scheme of the proposed neural network model for speech enhancement.

average method [4]. This is followed by the first processing block responsible for the transforming of the spectral decomposition of the entire frame (wideband processing) and subsequent non-linear ReLU transformation [13]. The resulting F-component vector G_t is partitioned into separate bands of the N_G hemisphere using cyclic index partitioning. For each central frequency band f, indices were chosen from (f-N_G) mod F to (f + N_G) mod F, where the operation of taking the remainder when dividing by the number of frequency bands F is indicated as mod. A similar operation of band allocation in the spectrum is applied to the input real vectors of the amplitudes X(t, f). The half-width of the frequency bands is N_X here. Bands corresponding to each other at the central frequency f are concatenated and normalized by dividing by the average value of the previously concatenated vectors for this frequency band. The calculated normalized subband representation of the input signal is fed to the second processing block (subband processing). This block is followed by a linear layer with an output of size F $\times$ 2 corresponding to the computed compressed complex mask $\mathrm{Cmpr}(M_{t-\tau}(t, f))$.

The mean square error (MSE) on the values of the compressed masks is used as the loss function in all experiments:

$$\mathcal{L}(M_{cl}, M) = \mathrm{MSE}(\mathrm{Cmpr}(M_{cl}), \mathrm{Cmpr}(M)), \tag{5}$$

where $M_{cl} = X/S$ is the true mask of the ratio of clean and distorted signals calculated for this training example.

2.3 Types of Processing Units

The most common type of transform used in processing sequential data are Long Short-Term Memory (LSTM) cells [13]. In a compact form, the LSTM transformation can be written as:

$$c_k = \sigma(W_f[h_{k-1}|x_k])c_{k-1} + \sigma(W_i[h_{k-1}|x_k])\tanh(W_c[h_{k-1}|x_k]) \tag{6}$$

$$h_k = \sigma(W_o[h_{k-1}|x_k])\tanh(c_k) \tag{7}$$

where $x_k \in R^{d_{in}}$ is the k-th vector of the input sequence, $h_{k-1}, h_k \in R^{d_h}$ are the previous and current vectors of the hidden state (simultaneously playing the role of output values),

$c_{k-1}, c_k \in R^{d_h}$ are the previous and current state vectors of the LSTM cell, $\sigma(\bullet)$ is the sigmoid function, $\tanh(\bullet)$ is the hyperbolic tangent function, | is the vector concatenation operator. The first term on the right-hand side of Eq. (6) is responsible for forgetting unnecessary information stored in the LSTM cell, the second for adding new information from the input data and the processed hidden state. In Eq. (7), a new hidden state (and the output of the cell) is formed based on the calculated state of the cell.

State Space for Sequence Modelling (S4)-transformation [6] in discrete form is described by the following formulas:

$$h_k = e^{A\Delta} h_{k-1} + \Delta B x_k, \tag{8}$$

$$y_k = C h_k + D x_k, \tag{9}$$

where $x_k \in R^{d_{in}}$ is the k-th vector of the input sequence, $h_{k-1}, h_k \in R^{d_{state}}$ are the previous and current state space vectors, $y_k \in R^{d_{in}}$ is the output sequence vector, $A \in R^{d_{in} \times d_{state}}$ is a trainable transition matrix for the hidden state space, and $B \in R^{d_{in} \times d_{state}}$, $C \in R^{d_{in} \times d_{state}}$ are two trainable projection matrices of input vectors into the internal hidden representation and from the hidden to the external output representation, respectively, $D \in R^{d_{in} \times d_{in}}$ – the trainable matrix of occurrence of the input representation in the output, Δ – the sampling step is a constant value. In essence, the S4-transformation can be seen as applying an input-controlled filter to the internal state and computing based on the last output vector.

The Mamba transformation [7] is based on a linear model of the hidden state space in discrete form, except for the addition of selectivity (control of the S4-transformation parameters):

$$h_k = e^{A\Delta} h_{k-1} + \Delta B(x_k) x_k = e^{A\Delta} h_{k-1} + \Delta (x_k W_B) x_k \tag{10}$$

$$y_k = C(x_k) h_k + D x_k = \Delta (x_k W_C) h_k + D x_k, \tag{11}$$

$$\Delta = Softplus[x_k W_{\Delta_1} W_{\Delta_2} + \Delta_{bias}], \tag{12}$$

where all notations in Eqs. (10) and (11) correspond to those in the S4-transfor-mation, but linear functions of the input sequence vector $B(x_k) = x_k W_B$, $C(x_k) = x_k W_C$ are used, defining dynamically varying projection matrices with trainable matrices $W_B \in R^{d_{in} \times d_{state}}$, $W_C \in R^{d_{in} \times d_{state}}$. The trainable matrices $W_{\Delta_1}, W_{\Delta_2}$ and the vector Δ_{bias} determine the dynamic sampling coefficient, which allows you to control the degree of "focus" on the current input signal [8]. The coefficients in expressions (8) and (9) are replaced by dynamic ones depending on the input data.

Table 1 shows the hyperparameters of the LSTM, S4 and Mamba layers used in the full-band and sub-band processing blocks. The dimensions of the internal states of each of the blocks were the same.

Table 1. Hyperparameters of the processing blocks.

Type	Hyperparameter	Value
LSTM	N_{layers}	2
	Wideband d_h	512
	Subband d_h	384
S4	N_{layers}	3
	d_{state}	32
	Wideband d_{in}	512
	Subband d_{in}	384
Mamba	N_{layers}	4
	d_{state}	16
	Wideband d_{in}	512
	Subband d_{in}	384

3 Experiments and Results

3.1 Dataset Description

All examples, for both training and testing, were generated from clean English audio recordings and noise examples from the DNS Challenge dataset [1]. The clean signals in this set consisted of 500 h of speech recordings from 2150 speakers selected from the LibriSpeech set [14]. The selection criterion for inclusion in the DNS Challenge set was the value of the MOS metric. Samples with a MOS greater than 4.0 were selected. Noise was added using 10-s recordings representing one of 150 classes of noise audio, with each noise class represented by 500 samples. The clean recording was combined with a randomly selected noise recording. In this case, the samples from the noise recording were multiplied by a coefficient chosen so that the signal-to-noise ratio in the resulting sample was a predetermined value.

When generating a training sample, the values of a given SNR were randomly selected from a uniform distribution in the range of values 0–40 dB. Two types of the generated audio were used for testing. The first type of audio samples was taken directly from the DNS Challenge. They were generated by the organizers of the DNS Challenge and contained approximately 150 samples with random SNRs in the range 0–20 dB. They are used in this paper for the final comparison of models with different types of transformation blocks and other best alternative models. The second type samples were generated by the authors specifically for this work. The signal-to-noise ratio for them was not random, but 6 subsets of 1000 samples were generated, each with an SNR value from −5 dB to 20 dB in increments of 5 dB. This set of audio samples is necessary to evaluate and compare how conventional models perform individually on existing distorted (-5–0 dB), moderately distorted (5–10 dB) and relatively mildly distorted (15–20 dB) audio signals.

3.2 Implementation Details

The conditions of numerical experiments on neural network training were generally consistent with previous works [11, 12]. The PyTorch deep learning library was used to implement the proposed model in Python. A workstation with a Nvidia GeForce 3090 GPU card with 24 GB VRAM was used in the training process. The number of processed audio examples in one batch was 4 pieces and their duration was 3 s. Adam optimizer with a learning rate of 10–4 was used. The STFT transform was computed on 32 ms frames with 50% overlap. The mask computation delay $\tau = 2$ frames.

3.3 Evaluation Metrics

The evaluation of the quality of the neural network versions for speech enhancement was carried out using several metrics to assess the quality of speech audio examples. According to the P.808 standard [15], the Mean Opinion Score (MOS) is estimated as the average of the overall signal quality estimates of 20 to 60 people. The evaluation signals are adjusted to the same volume level and may be repeated in several independent evaluation sessions. Each listening session ends with a score on a scale from 1 (unacceptable quality) to 5 (perfect quality). The MOS value of 4.0 and above is considered to correspond to a high quality speech signal. In this paper, it was not possible to organize sessions for human rating of speech signal quality, but automatic rating methods [16, 17, 19] were used, which in one way or another approximate the rating of human listeners.

The first pair of evaluation methods was the calculation of the Perceptual Evaluation of Speech Quality (PESQ) [16], which provides a score on a scale from 0.5 (worst quality) to 4.5 (ideal quality) for a pair of reference clean and appropriately controlled (distorted or enhanced) speech fragments. There are two versions of PESQ: narrowband for a telephone channel with a sampling frequency of 8 kHz (NB PESQ) and wideband for multimedia/video conferencing with a sampling frequency of 16 kHz (WB PESQ).

The second metric was the Short-Term Objective Intelligibility (STOI) [17], measured as a percentage. It was assessed using the reference clean and controlled (noisy or enhanced) resampled signals at a sampling rate of 10 kHz.

The third group of metrics simulated a triple assessment of the quality of speech fragments using the methodology of the P.835 standard [18] using a pre-trained deep neural network DNSMOS [19]. The main advantage of this estimation method was the absence of the need for an exemplary clean speech signal, which made it possible to build distributions of this metric for all subsets of speech data used in the work. The triple quality score consisted of values denoted by SIG (exclusive assessment of the degree of speech distortion in the signal), BAK (degree of perceptibility of background noise) and OVRL (general assessment of the quality of the speech signal, similar to P.808).

3.4 Comparison of Different Network Versions for Various Noise Level

Table 2 shows a comparison of the quality of work on all the considered metrics for three versions of the neural network architecture – based on LSTM, S4 and Mamba transformations. Each of the models was trained on the same training set generated based on samples from the DNS Challenge. For the initial quality assessment and comparison of the three models, the same test samples with preset noise levels from -5 dB to 20 dB, described above, were used.

Table 2. Performance of the proposed models for signals with different SNR.

SNR, dB	Model	WB PESQ	NB PESQ	STOI	DNSMOS SIG	DNSMOS BAK	DNSMOS OVRL
−5	noisy	1.1229	1.3681	0.4722	1.7836	1.4477	1.3949
	LSTM	**1.5725**	**2.0629**	**0.7706**	**3.2019**	**3.2101**	**2.4376**
	S4	1.4873	1.9593	0.7608	2.9088	3.0815	2.3332
	Mamba	1.5326	2.0155	0.7704	2.9732	3.0928	2.3832
0	noisy	1.1766	1.5371	0.5970	2.3076	1.7179	1.6637
	LSTM	**1.9342**	**2.5200**	0.8759	**3.1717**	**3.5213**	**2.702**
	S4	1.8366	2.4243	0.8677	3.1337	3.4367	2.6348
	Mamba	1.9088	2.4943	**0.8790**	3.1656	3.4493	2.67
5	noisy	1.2726	1.7424	0.6954	2.9239	2.0676	2.1976
	LSTM	**2.3506**	**2.9199**	0.9255	**3.295**	**3.7359**	**2.9001**
	S4	2.2405	2.8341	0.9243	3.2584	3.681	2.845
	Mamba	2.3335	2.9122	**0.9260**	**3.295**	3.713	2.8821
10	noisy	1.4686	2.0313	0.7788	3.2781	2.4582	2.3223
	LSTM	**2.7886**	3.3113	0.9579	3.3808	**3.855**	**3.0294**
	S4	2.6835	3.2533	0.9561	3.354	3.8224	2.9919
	Mamba	2.7753	**3.3156**	**0.9581**	**3.3909**	3.8511	3.0205
15	noisy	1.7781	2.4134	0.8366	3.4539	2.8761	2.6015
	LSTM	3.2239	3.6501	0.9764	3.4444	**3.9491**	**3.1263**
	S4	3.1398	3.6012	0.9755	3.4247	3.9298	3.1028
	Mamba	**3.2291**	**3.6621**	**0.9765**	**3.4451**	3.9459	3.1186
20	noisy	2.2578	2.8572	0.8810	3.4902	3.1909	2.7853
	LSTM	3.5592	3.8988	0.9835	3.4678	**3.9658**	3.1603
	S4	3.5026	3.8669	0.9846	3.4527	3.9493	3.1419
	Mamba	**3.5628**	**3.9120**	**0.9850**	**3.4686**	3.9634	**3.1634**

The main metrics used to evaluate the trained models were DNSMOS estimates [19], as they are closest to what human experts can give. Figures 2, 3 and 4 show the dependence of the average DNSMOS SIG, BAK and OVRL values on a given signal-to-noise ratio of the test examples. It can be seen that for highly distorted files with a level of SNR = −5 dB, the model with LSTM transformations performed best in all three metrics. Note that there were no examples with similar SNR values in the training sample. The high results of the recurrent network can be attributed to both the best generalization ability and the fact that the number of parameters in the transformation blocks for it was the largest of all three versions of the network. The most significant gap was according to the BAK metric, i.e. the network suppressed background noise quite well. When the SNR was increased to 0–10 dB, most quality metrics differed slightly for all models, but the Mamba model gradually outperformed the others. Finally, at SNR = 15–20 dB, corresponding to less noticeable background noise, the Mamba model became the best, ahead of all others in all metrics except BAK (see Table 2). A neural network based on the S4 transformation was always in second or third place in the assessment of the quality of the speech enhancement results, probably due to a low generalization ability and/or a lower ability to model complex non-linear transformations necessary for effective noise suppression.

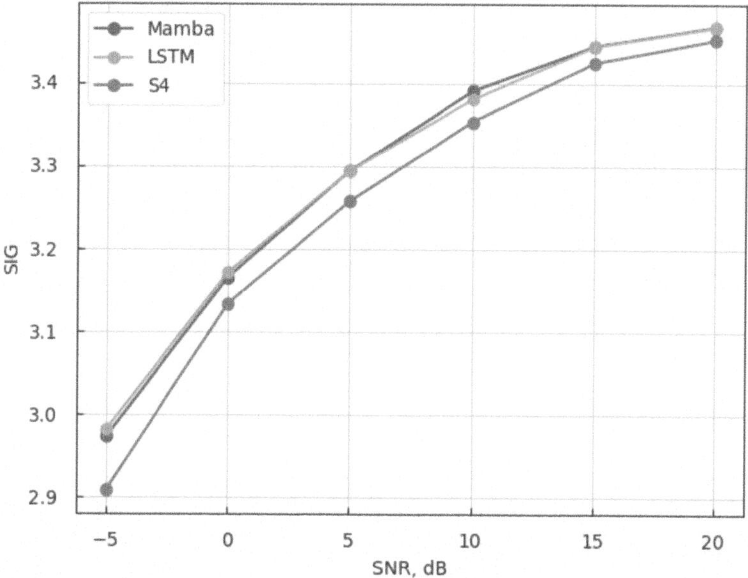

Fig. 2. Comparison of networks with different types of transformations according to the DNSMOS SIG metric.

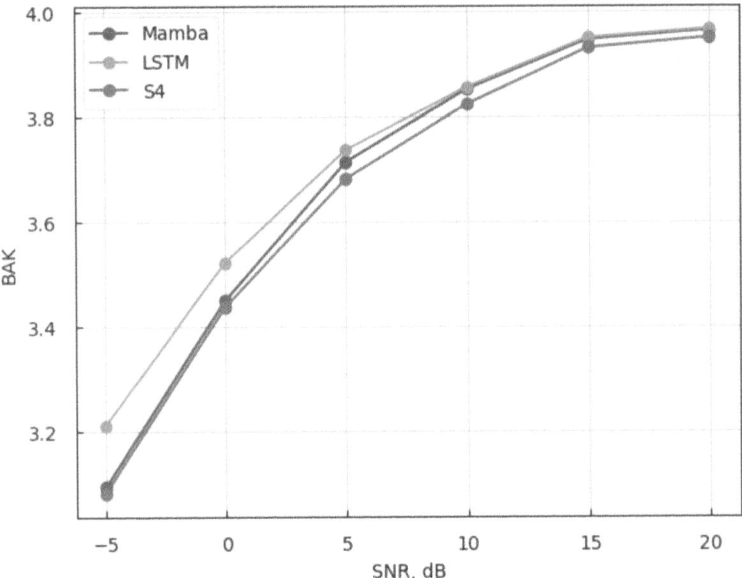

Fig. 3. Comparison of networks with different types of transformations according to the DNSMOS BAK metric.

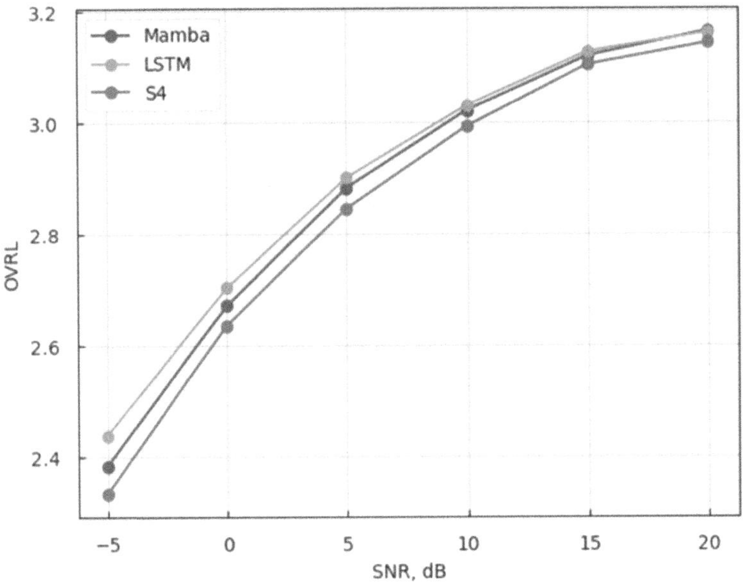

Fig. 4. Comparison of networks with different types of transformations according to the DNSMOS OVRL metric.

3.5 Comparison with Alternative Solutions

The performance of the proposed modelы шs summarized in Table 3 along with a comparison with the best alternative models. The comparison was performed on 150 samples of the DNS Challenge test subset. The higher the metrics, the higher the quality of the model. The benchmark values of the quality metrics were calculated for the unimproved noise signals in the top row, which were labelled as 'Noisy signals'. The versions of the quality improvement models considered in this paper were compared with each other and with two other state-of-the-art models that are designed for efficient speech enhancement. The first was the DCCRN-E model based on an architecture combining complex convolutional layers and complex recurrent transformations [4]. The second model was Conv-TasNet [20], which was a convolutional end-to-end network with low latency of the processed signal, operating on raw audio stream. The comparison of the models showed that the Mamba-based model is the most optimal solution in terms of the quality of signal enhancement and the number of parameters to be trained.

Table 3. Comparison of the proposed models on the DNS Challenge test set.

Model	Model size, $\times 10^6$	NB-PESQ	WB-PESQ	STOI
Noisy signals	–	2.454	1.582	91.52
TS-LSTM	7.5	3.338	2.832	96.05
Conv-TasNet	3.7	3.266	–	–
LSTM-based	5.62	3.305	2.777	96.11
S4-based	4.80	3.295	2.767	96.53
Mamba-based	1.19	3.374	2.853	96.31

4 Conclusion

In this paper, the quality of work of three versions of a neural network for speech enhancement based on various types of sequential data transformations is compared. It was shown that the model based on the Mamba-transformation turned out to be the best in a wide range of test conditions. It surpassed not only other trained versions of the networks from this paper, but also alternative methods of cleaning from distortion. This neural network model was distinguished not only by the high quality of the output signal, but also by a relatively small number of trainable parameters, which makes it promising for building new signal processing methods on devices with limited computing abilities.

Acknowledgements. The study was carried out within the framework of the project "Development of the Automatic Expert System for Identifying Illegal Modifications of Audio Files" as a part of the Altai State University strategic development program within the "Priority 2030" state academic leadership program (2021–2030).

References

1. Reddy, C., Beyrami, E., Dubey, H.: Deep noise suppression challenge: Datasets, subjective testing framework, and challenge results. In: Proceedings of INTERSPEECH 2020, 25–29 October 2020, pp. 2492–2496. Shanghai, China (2020)
2. Dubey, H., Aazami, A., Gopal, V., Naderi, B., Braun, S., Cutler, S.: ICASSP 2023 deep noise suppression challenge. IEEE Open J. Signal Process. **5**, 725–737 (2024)
3. Hao, X., Su, X., Horaud, R, Li, X.: FullSubNet: a full-band and sub-band fusion model for real-time single-channel speech enhancement. In: IEEE International Conference on Acoustics, Speech, and Signal Processing (ICASSP), 6–11 June 2021, pp. 6633–6637. Toronto, Canada (2021)
4. Hu, Y., et al.: DCCRN: deep complex convolution recurrent network for phase-aware speech enhancement. In: Proceedings of INTERSPEECH 2020, 25–29 October 2020, pp. 2472–2476. Shanghai, China (2020)
5. Umut, I., Giri, R., Phansalkar, N., Valin, J.-M., Helwani, K., Krishnaswamy, A.: PoCoNet: better speech enhancement with frequency-positional embeddings, semi-supervised conversational data, and biased loss. In: Proceedings of INTERSPEECH 2020, 25–29 October 2020, pp. 2487–2491. Shanghai, China (2020)
6. Gu, A., Goel, K., Ré, Ch.: Efficiently modeling long sequences with structured state spaces. In: Proceedings of The Tenth International Conference on Learning Representations, ICLR 2022, Virtual Event, 25–29 April 2022, (2022)
7. Gu, A., Dao, T.: Mamba: linear-time sequence modeling with selective state spaces. ArXiv preprint, arXiv: 2312.00752, 37 pp (2023)
8. Du, Y., Liu, X., Chua, Y.: Spiking structured state space model for monaural speech enhancement. In: ICASSP 2024 - 2024 IEEE International Conference on Acoustics, Speech and Signal Processing (ICASSP), 14–19 April 2024, pp. 766–770. Seoul, Republic of Korea (2024)
9. Loizou, P.C.: Speech enhancement: theory and practice. CRC Press, Boca Raton, FL, USA (2007)
10. Williamson, D.S., Wang, Y., Wang, D.: Complex ratio masking for monaural speech separation. IEEE/ACM Trans. Audio, Speech Lang. Process. **24**(3), 483–492 (2016)
11. Nasretdinov, R., Ilyashenko, I., Lependin, A.: Two-stage method of speech denoising by long short-term memory neural network. In: HPCST 2021: High-Performance Computing Systems and Technologies in Scientific Research, Automation of Control and Production, Communications in Computer and Information Science, vol. 1526, pp. 86–97. Springer (2022)
12. Lependin, A., Karev, V., Nasretdinov, R., Ilyashenko, I.: Speech enhancement based on two-stage neural network with structured state space for sequence transformation. In: HPCST 2023: High-Performance Computing Systems and Technologies in Scientific Research, Automation of Control and Production, Communications in Computer and Information Science, vol. 1986, pp. 16–28. Springer (2024)
13. Goodfellow, I., Bengio, Y., Courville, A.: Deep learning. MIT Press, Cambridge, MA, USA (2016)
14. Panayotov, V., Chen, G., Povey, D., Khudanpur, S.: Librispeech: An ASR corpus based on public domain audio books. In: 2015 IEEE International Conference on Acoustics, Speech and Signal Processing (ICASSP), 19–24 April 2015, pp. 5206–5210. South Brisbane, QLD, Australia (2015)
15. International Telecommunication Union: ITU-T P.808 Subjective evaluation of speech quality with a crowdsourcing approach (2018)

16. Rix, A.W., Beerends, J.G., Hollier, M.P. Hekstra, A.P.: Perceptual evaluation of speech quality (PESQ) – a new method for speech quality assessment of telephone networks and codecs. In: IEEE International Conference on Acoustics, Speech, and Signal Processing. Proceedings, 7–11 May 2001, pp. 749–752. Salt Lake City, UT, USA (2001)
17. Taal, C.H., Hendriks, R.C., Heusdens, R., Jensen, J.: A short-time objective intelligibility measure for time-frequency weighted noisy speech. In: IEEE International Conference on Acoustics, Speech and Signal Processing, 15–19 March 2010, pp. 4214–4217. Dallas, TX, USA (2010)
18. International Telecommunication Union: ITU-T P.835 Subjective test methodology for evaluating speech communication systems that include noise suppression algorithm (2011)
19. Reddy, C.K.A., Gopal, V., Cutler, R.: Dnsmos P.835: a non-intrusive perceptual objective speech quality metric to evaluate noise suppressors. In: ICASSP 2022 - 2022 IEEE International Conference on Acoustics, Speech and Signal Processing (ICASSP), 23–27 May 2022, vol. 2022, pp. 886–890. Singapore, Singapore (2022)
20. Luo, Y., Mesgarani, N.: Conv-TasNet: surpassing ideal time-frequency magnitude masking for speech separation. IEEE/ACM Trans. Audio, Speech Lang. Process. **27**(8), 1256–1266 (2019)

Locating the Specific Melodic Constructs in a Midi Files Library Using Digital Audio Fingerprints

Pavel Ladygin(✉), Alexander Mansurov, and Andrey Lependin

AltSU – Altai State University, Lenin Avenue 61, 656049 Barnaul, Russia
ladyginps@mc.asu.ru

Abstract. The paper presents a novel approach to analyzing and comparing melodic constructs in MIDI format. The calculated digital audio fingerprints of these constructs contain the encoded transitions between note pitches and distances between their heights. It helps identify and compare melodic constructs with their variety of possible interpretations played with different chords, in different keys, tempos, etc. This approach can be useful for searching the music in the music library or helpful for music composers to avoid unintended plagiarism. There are also many prospects when using the approach of real forensic studies and the tasks of verification and authentication of music pieces and recordings. The algorithm developed within the proposed approach includes the required steps and procedures to process music compositions, calculate their digital audio fingerprints, and conduct the search of the specific music composition using the audio fingerprint data. Performance testing is done using the arranged data set, which includes 894 MIDI files of various music compositions played with different instruments and music effects. The obtained results demonstrate the effectiveness and robustness of the developed algorithm and the proposed approach to matching audio fingerprints and identifying similar music compositions with 100% accuracy. Additionally, the concept of using the developed algorithm to locate music in online music libraries by "humming" is presented and tested in the paper.

Keywords: Digital Fingerprint · Music Audio Signal · MIDI · Audio Recording Authentication · Expert Assessment

1 Introduction

Digital technologies have already become an essential part of world music art when it comes to the creation, adaptation, and distribution of musical compositions. Existing online music libraries and collections of audio files provide tools for searching the desired music by names and author names, as well as music genres and user preferences. In 2020, Google announced the novel "Hum to Search" feature, allowing users to locate a song simply by humming it. In 2023, this feature was introduced by YouTube, although in an experimental stage. Algorithms helping to find the specific melodic constructs by "humming" it or providing a sample can be useful not only for ordinary users but for

V. Jordan et al. (Eds.): HPCST 2024, CCIS 2919, pp. 28–38, 2026.
https://doi.org/10.1007/978-3-032-20325-0_3

music composers as well to avoid the unnecessary problems and unintended plagiarism. Thus, it is necessary to develop new methods and software algorithms to analyze and compare music audio files.

The concept of locating the desired melody by "humming" was already introduced in 2000 [1] for MIDI audio files (Fig. 1).

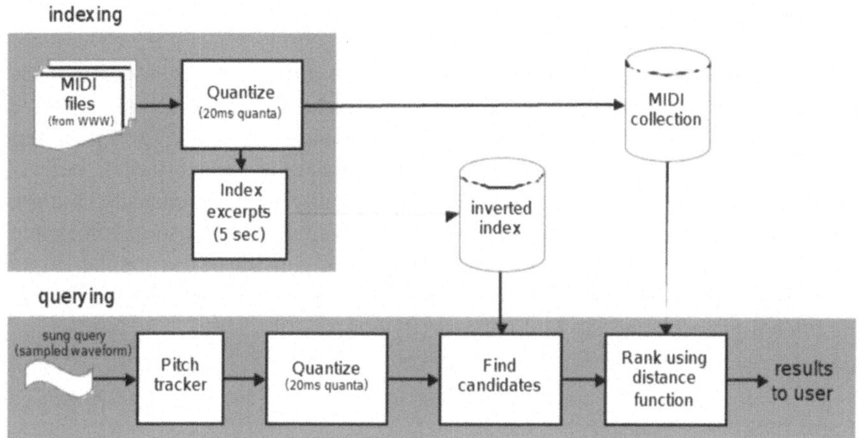

Fig. 1. Digital library of popular music: indexing and processing search queries.

Authors of the paper [1] proposed splitting the "humming" voice audio sequence into 20-ms parts to analyze each of them further with a pitch tracker and locate the frequencies of the root positions. Then, the obtained tuple can be used for searching through the library of MIDI files for matches. The choice of MIDI format was explained by its popularity among the music makers, less memory usage, and sufficient capability to represent melodic constructs by discretizing each music note with its pitch, length, and intensity. And it's exactly these characteristics that represents the music composition accumulatively are of great interest when following copyright laws dealing with the musical compositions. However, the proposed approach does not consider aspects of tonal matching (music played in different keys) and tempo, which limits the effectiveness of its application scope.

There are many software algorithms and special online services providing the music content identification [2–4] by analyzing and comparing the audio file data using spectrum analysis [5, 6], cepstral analysis [7], digital fingerprinting, etc. Almost all the algorithms that compare melodic constructs utilize the aspects of finding the timing patterns (or synchronism), automatic optimization of MIDI files data adjustment using the DTW (Dynamic Time Warping) [9] approach, comparison to find syntactic similarity of MIDI files using hashes [10], or using the 'thermal maps' of distribution of notes by channels [11]. Also, there are cases of direct comparison of melodic compositions using fuzzy matching with the Levenstein metrics [12].

The development and unification of the best approaches can produce effective methods, algorithms, and tools for locating music by "humming" it in many music libraries and collections.

2 The General Concept of the Proposed Method

2.1 MIDI Data Processing

The MIDI format is a very convenient way to store music, where its common data sequence follows the flow of the actual music. A typical MIDI message looks like this example: message('type', channel = X, note = Y, velocity = Z, time = Q), where 'type' can be 'note_on' or 'note_off', 'channel' – is the number of a channel (0–15), 'note' – is the number of a note within the range (0–127), 'velocity' – is the intensity of a pitch (0–127), and 'time' – is the length of a note in 'ticks'. Figure 2 demonstrates an example of the MIDI data obtained by processing the "hummed" version of the popular Russian song "A Spruce Was Born in the Forest" (composed by M. Krasev) with the Spotify's Basic Pitch [7] service and the Mido library (Python).

```
MidiFile(type=1, ticks_per_beat=480, tracks=[
  MidiTrack([
    MetaMessage('track_name', name='', time=0),
    MetaMessage('set_tempo', tempo=500000, time=0),
    MetaMessage('end_of_track', time=0)]),
  MidiTrack([
    MetaMessage('track_name', name='', time=0),
    Message('program_change', channel=0, program=4, time=0),
    Message('note_on', channel=0, note=79, velocity=87, time=0),
    Message('note_off', channel=0, note=79, velocity=0, time=624),
    Message('note_on', channel=0, note=76, velocity=111, time=22),
    Message('note_off', channel=0, note=76, velocity=0, time=312),
    Message('note_on', channel=0, note=76, velocity=96, time=1),
    Message('note_on', channel=0, note=79, velocity=84, time=323),
    Message('note_off', channel=0, note=76, velocity=0, time=101),
    Message('note_off', channel=0, note=79, velocity=0, time=524),
    Message('note_on', channel=0, note=76, velocity=110, time=0),
    Message('note_off', channel=0, note=76, velocity=0, time=334),
    Message('note_on', channel=0, note=76, velocity=97, time=0),
    Message('note_on', channel=0, note=79, velocity=79, time=313),
    Message('note_off', channel=0, note=76, velocity=0, time=99),
    Message('note_off', channel=0, note=79, velocity=0, time=213),
    Message('note_on', channel=0, note=77, velocity=103, time=11),
    Message('note_off', channel=0, note=77, velocity=0, time=334),
    Message('note_on', channel=0, note=76, velocity=98, time=0),
    Message('note_on', channel=0, note=74, velocity=97, time=312),
    Message('note_off', channel=0, note=76, velocity=0, time=45),
    Message('note_on', channel=0, note=72, velocity=99, time=269),
    Message('note_off', channel=0, note=74, velocity=0, time=22),
    Message('note_off', channel=0, note=72, velocity=0, time=480),
    Message('note_on', channel=0, note=65, velocity=0, time=902),
    Message('note_off', channel=0, note=65, velocity=0, time=11),
    MetaMessage('end_of_track', time=0)])
```

Fig. 2. An example of MIDI data.

2.2 The Proposed Processing Algorithm

It seems possible to calculate the digital fingerprint of the music composition with its MIDI data using the algorithm [13] developed earlier and proven to be effective when processing wav-files. The algorithm [13] encodes transitions between note pitches into binary sequences. Its further improvement by the authors and direct usage of the MIDI data provide new opportunities to calculate and produce more complete and precise binary sequences of melodic constructs. The updated version of the algorithm [13] proposed in the paper processes MIDI data and encodes transitions between note pitches as follows: "01" – when pitch height increases, "10" – when pitch height decreases, "11" – when pitch height remains unchanged. Additionally, an absolute distance between the heights of note pitches in transitions is calculated in all MIDI channels using values of the 'note' parameter. Figure 3 presents the simplified encoding and calculation scheme for a MIDI channel of the processed MIDI data.

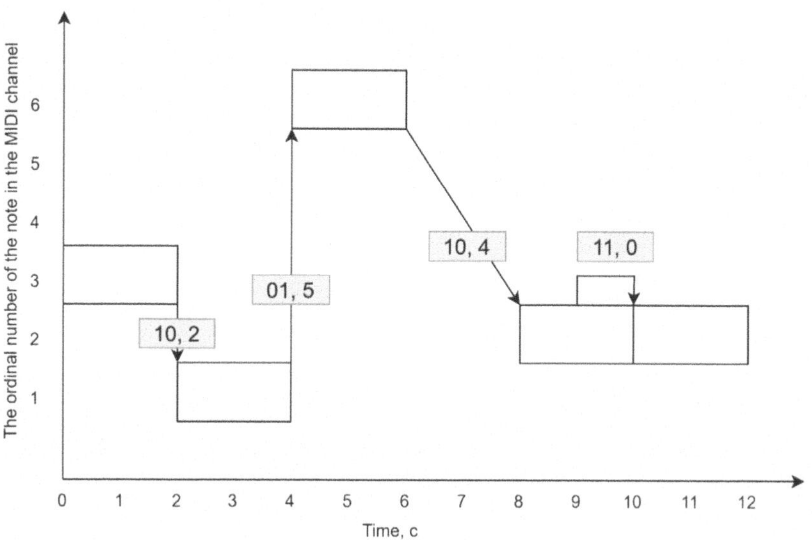

Fig. 3. Scheme of encoding transitions between note pitches and calculating the absolute distance.

The processing of music chords requires additional consideration. Typically, the root note in a chord should be the most intense one. The proposed algorithm follows this assumption by comparing the values of the 'velocity' parameter. The note with the highest 'velocity' should be the one to be used when encoding the transitions and calculating the absolute distance. The first note is used if the 'velocity' values of two or more notes are equal.

Any musical composition can be treated as a number of combined and properly arranged melodic constructs. Any melodic construct has artistic value and enriches the musical composition. Since the MIDI format supports up to 16 channels, they can all contain the data of melodic constructs. Therefore, all MIDI channels should be processed to produce the digital fingerprint.

The general steps of the proposed algorithm to calculate the digital fingerprint of a MIDI file is presented in Fig. 4.

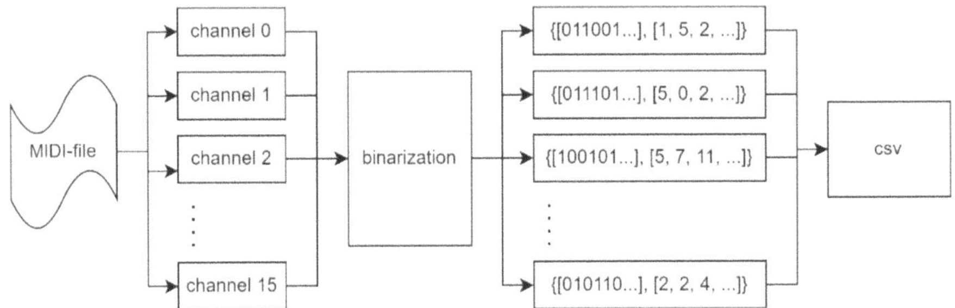

Fig. 4. General steps of the proposed algorithm to calculate the digital fingerprint of a MIDI file.

The processed results are saved step by step into a CSV file in a following format: 'Filename', 'Channel Number', 'Tuple of Encoded Transitions', 'Tuple of Calculated Distances'. Here, the CSV file data form the digital fingerprint of a music composition that is based only on arrangement of notes (or pitches) and independent to music keys and tempo. This should be helpful to find the correct audio file of the original music composition in the audio library using the "hummed" sample because, in most cases, the "hummed" melody tends to differ in key, note pitches, and tempo from the original music composition.

3 Experiments and Results

3.1 Data Set for Testing

The arranged data set for testing the proposed algorithm consists of 894 MIDI files selected from the publicly available Tegridy MIDI data set. Each MIDI file contains a single music composition. All MIDI files utilize up to 16 channels, and their time lengths vary from 4 s up to 210 s. There is a wide range of musical instruments in the selected music compositions, including string, keyboard, wind, and percussion sets.

Recordings of five different Russian songs are used as samples for searching and matching the music compositions in the arranged data set:

- Song 1 - "A Spruce Was Born in the Forest" (composed by M. Krasev)
- Song 2 – "Cheburashka's Song" (composed by V. Shainsky)
- Song 3 – "Magic Land" (composed by V. Shainsky)
- Song 4 – "The Green Grasshopper" (composed by V. Shainsky)
- Song 5 – "The Dog is Biting" (composed by S. Nikitin)

Two versions, "hummed" and played by guitar, of each song are recorded using the built-in microphone of the Lenovo IdeaPad Gaming 3 15IMH05 laptop. The recordings are further processed by the Spotify's Basic Pitch service and converted into ten single-channel MIDI files with a sequence of 9–12 notes in each file. Visualizations of the MIDI files (Figs. 5 and 6) demonstrate the obvious shortcomings of such conversion, thus bringing the possible "real life" imperfections. The MIDI file of the "hummed" version of the first song is also added to the arranged data set to test the basic effectiveness and credibility of the proposed algorithm. The song "The Green Grasshopper" (Song 4) is chosen as a sample because there is a MIDI file of this song played by a set of instruments in the arranged data set.

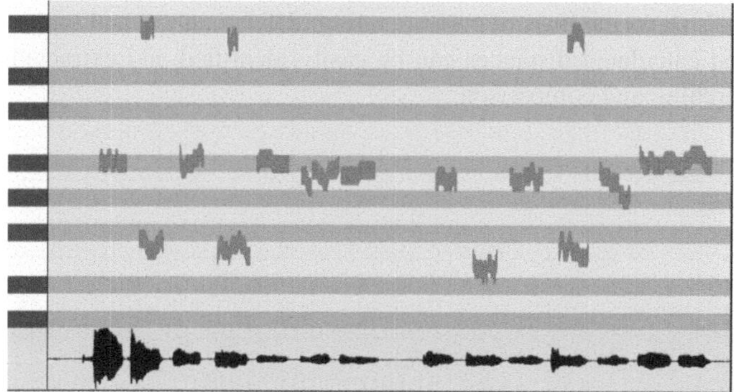

Fig. 5. Visualization of the "hummed" Song 4.

Fig. 6. Visualization of the Song 5 played by guitar.

Melodic constructs visualized in Figs. 5 and 6 contain obvious gaps between note pitches, vibrato elements and the simultaneous playing of two notes, even when the song is "hummed". The proposed algorithm ignores these shortcomings, but they should be considered and dealt with in the future studies.

3.2 Implementation Details

The proposed algorithm is implemented using the Python language, with the Mido library to process MIDI files. The developed programming code includes all the required procedures for the algorithm steps mentioned earlier. Also, there are additional procedures to visualize the processed data, search and compare the digital fingerprint data in an array of CSV files obtained after processing music compositions from the data set.

A typical example of digital fingerprint data processed and saved in a CSV file is shown in Fig. 7. Here, the lines of corresponding MIDI channels and tuples of encoded transitions and calculated distances can be easily identified and extracted for further actions of searching and comparison.

```
Tegridy-Children-Songs-046-CC-BY-NC-SA.mid|0|['011010011001100110100110011001010110011111101101001011001
Tegridy-Children-Songs-046-CC-BY-NC-SA.mid|1|['010111100110011110100111101011011010110101100101100110011O
Tegridy-Children-Songs-046-CC-BY-NC-SA.mid|4|['0110100110110111011110110111011110110101011001100110010101
Tegridy-Children-Songs-046-CC-BY-NC-SA.mid|5|['110111011101111011011110111011011110110111101101111011011
Tegridy-Children-Songs-046-CC-BY-NC-SA.mid|6|['1011110101']|[[17, 0, 0, 2, 7]]
Tegridy-Children-Songs-046-CC-BY-NC-SA.mid|8|['1101101010100111100110100']|[[0, 1, 1, 2, 12, 9, 28, 0, 12, 7, 20, 3]]
Tegridy-Children-Songs-046-CC-BY-NC-SA.mid|9|['1011111111100111111111111111111111111001111111111111110011111
Tegridy-Children-Songs-046-CC-BY-NC-SA.mid|11|['']|[[]]
Tegridy-Children-Songs-047-CC-BY-NC-SA.mid|0|['111111111001011111111111010111111100101100110111111111111001
Tegridy-Children-Songs-047-CC-BY-NC-SA.mid|2|['1001011001101001011010010101100101011110101010011001011001
Tegridy-Children-Songs-047-CC-BY-NC-SA.mid|4|['1001']|[[3, 2]]
Tegridy-Children-Songs-047-CC-BY-NC-SA.mid|5|['0101010101100110011001101001010101011010100101100110100110
Tegridy-Children-Songs-047-CC-BY-NC-SA.mid|6|['0110011001011001100110101001011010010101101001101010111001
Tegridy-Children-Songs-047-CC-BY-NC-SA.mid|7|['1010110110010110100110011010100111011010111001011010011010
Tegridy-Children-Songs-047-CC-BY-NC-SA.mid|8|['1110011001101001111001100111100110011001101001111']|[[0, 2, 2, 3
Tegridy-Children-Songs-047-CC-BY-NC-SA.mid|11|['1001101001100110100110100110100110011010010110100110100101
Tegridy-Children-Songs-047-CC-BY-NC-SA.mid|12|['0101111011101001011001011110101101111010010101011101101001
Tegridy-Children-Songs-047-CC-BY-NC-SA.mid|13|['100110110110100110010111100110011001101101101010100101111
Tegridy-Children-Songs-048-CC-BY-NC-SA.mid|0|['111111111101100101100101101110011011111111111111001100101111
Tegridy-Children-Songs-048-CC-BY-NC-SA.mid|1|['1101110111101110011001110111101101111011100101010101011011
Tegridy-Children-Songs-048-CC-BY-NC-SA.mid|2|['1011101110111011101110111011101110110110011010011010011011
Tegridy-Children-Songs-048-CC-BY-NC-SA.mid|3|['1010111011101011011010111011100111011010111010011101100110
Tegridy-Children-Songs-048-CC-BY-NC-SA.mid|4|['11101110111011101110110111101110111011101101110111101110111011
Tegridy-Children-Songs-048-CC-BY-NC-SA.mid|5|['1111']|[[0, 0]]
```

Fig. 7. Example of a CSV file data fragment. Total number of lines – 3590, file size – 7 Mb.

Next, the converted MIDI file of the "hummed" melody is processed in a similar way to obtain the tuples of encoded transitions and calculated distances. They further become the input data (or the complex search key) for the procedures performing the search of a similar music composition. There are two steps for searching and comparison using the complex search key:

- Find the parts of the input tuple of encoded transitions (of the search key) matching as a substring the data of corresponding tuples in every obtained CSV file after processing the data set. Persistent matchings of encoded transitions found in a CSV file of a processed music composition set this composition (its digital fingerprint or CSV file data) as a candidate for the next step.
- Among the selected candidates, find and compare the parts of the input tuple of calculated distances (of the search key) with the data of the corresponding tuples in the

CSV files of the selected candidates. The total number of matched distances is calculated and used as a score to identify the best possible matches among the candidates. There should be at least eight matching distances to qualify the candidate as the best possible match. This score value falls under the assumption that a "hummed" piece should have at least nine notes or more.

The results of searching the arranged data set of music compositions for the "hummed" versions of songs 1–5 are shown in Table 1 and Fig. 8.

Table 1. Results of matching the "hummed" versions of songs.

N	Sample	Selected Candidates	Result
1	Song 1	3	Found
2	Song 2	0	Not found
3	Song 3	0	Not found
4	Song 4	0	Not found
5	Song 5	9	Not found

```
Is sublist present in list ? : True
Time start:  0
Channel:  0
The number of matching notes:  12
Filename:  elka.mid


Is sublist present in list ? : True
Time start:  0
Channel:  0
The number of matching notes:  10
Filename:  elka3.mid


Is sublist present in list ? : True
Time start:  62
Channel:  5
The number of matching notes:  9
Filename:  Gloryofl.mid
```

Fig. 8. Results of matching the "hummed" version of the Song 1.

The search results in Fig. 8 demonstrate that there are three music compositions (MIDI files) similar to the "hummed" one in the arranged data set. The details of the search state the following:

- elka.mid: this MIDI file is exactly the "hummed" music added to the arranged data set to test the ability of the proposed algorithm to find the exact identical music compositions;
- elka3.mid: this MIDI file contains the music composition of the "hummed" song played by the five music instruments (using five MIDI channels);
- Gloryofl.mid: this MIDI file contains a music composition that is completely different from the "hummed" song. However, the detailed analysis of its tuples data in the CSV file of this music composition reveals similarities in some parts of encoded transitions and calculated distances with the tuples data of the "hummed" song. This can bring new prospects for methods and techniques for analyzing melodic constructs.

The "hummed" sample of the Song 4 does not even produce the selected candidate from the first step of searching and matching. It can be explained by the converted MIDI file imperfections shown earlier and the lack of workaround procedures in the proposed algorithm. There are 9 candidates for the sample of Song 5, however no one among them is qualified as the best possible match. Playing all the MIDI files of the selected candidates and listening to the music confirms the obvious difference in music compositions.

The results of searching the arranged data set of music compositions for the versions of songs 1–5 played by guitar are shown in Table 2.

Table 2. Results of matching the guitar versions of songs.

N	Sample	Selected Candidates	Result
1	Song 1	3	Found
2	Song 2	2	Not found
3	Song 3	0	Not found
4	Song 4	3	Found
5	Song 5	0	Not found

Here, the guitar versions of the Song 1 and Song 4 are matched with the correct MIDI files of the same songs in the arranged data set. However, there are two mismatches for the Song 4 guitar sample, and they are proved to be completely different music compositions when playing the corresponding MIDI files and listening to them. The explanation of this fact is completely the same as in the case of the "hummed" version of Song 1 given earlier. It signifies that the algorithm requires further improvement to reduce the false matchings.

4 Conclusion

The paper proposes an algorithm for obtaining the digital audio fingerprints of music compositions using their converted MIDI versions. It employs a simple and convenient way to calculate digital audio fingerprints based on the peculiarities of note pitch transitions in melodic constructs. Obtained audio fingerprints can be easily matched and do not require sophisticated "visualizations" or further subjective or arbitrary interpretations. The methods and approaches incorporated in the proposed algorithm demonstrate its effectiveness and accuracy in matching the specific music composition played in different keys and tempos or using different music instruments. It also signifies the ability of the algorithm to identify a variety of possible interpretations of the same music composition. Despite the demonstrated advantages, the proposed algorithm requires additional improvement and more thorough testing and evaluation, which are planned in the future studies.

Acknowledgements. The study was carried out within the framework of the project "Development of the Automatic Expert System for Identifying Illegal Modifications of Audio Files" as a part of the Altai State University strategic development program within the "Priority 2030" state academic leadership program (2021–2030).

References

1. Francu, C., Nevill-Manning, C.G.: Distance metrics and indexing strategies for a digital library of popular music. In: Proceedings of the IEEE International Conference on Multimedia and Expo (ICME 2000), vol. 2, pp. 889–892. New York, NY, USA (2000). https://doi.org/10.1109/ICME.2000.871502
2. Nieuwenhuizen van, H.A., Venter, W.C., Grobler, L.M.: The study and implementation of Shazam's audio fingerprinting algorithm for advertisement identification. In: Proceedings of the SATNAC 2011 (2011)
3. Baluja, S., Covell, M.: Audio fingerprinting: Combining computer vision and data stream processing. In: Proceedings of the International Conference on Acoustics, Speech, and Signal Processing (ICASSP 2007), vol. 2, pp. II-213–II-216 (2007). https://doi.org/10.1109/ICASSP.2007.366210
4. Baluja, S., Covell, M.: Content fingerprinting using wavelets. In: Proceedings of the 3rd European Conference on Visual Media Production (CVMP 2006), pp. 198–207. London (2006). https://doi.org/10.1049/cp:20061964
5. Cano, P., Batlle, E., Kalker, T., Haitsma, J.: A review of audio fingerprinting. J. VLSI Signal Process. Syst. Signal Image Video Technol. **41**(3), 271–284 (2005). https://doi.org/10.1007/s11265-005-4151-3
6. Sonnleitner, R., Widmer, G.: Robust quad-based audio fingerprinting. IEEE/ACM Trans. Audio Speech Lang. Process. **24**(3), 409–421 (2016)
7. Bittner, R.M., Bosch, J.J., Rubinstein, D., Meseguer-Brocal, G., Ewert, S.: A lightweight instrument-agnostic model for polyphonic note transcription and multipitch estimation. In: Proceedings of the International Conference on Acoustics, Speech, and Signal Processing (ICASSP 2022), Singapore, pp. 781–785 (2022). https://doi.org/10.1109/ICASSP43922.2022.9746549
8. Kataeva, E.S., Yakimuk, A.Yu.: Application of synchronous isolation for evaluating the similarity of vocal performers. Proc. Tomsk State Univ. Control Syst. Radioelectron. **22**(3), 49–54 (2019) (in Russian)
9. Raffel, C.: Learning-Based Methods for Comparing Sequences, with Applications to Audio-to-MIDI Alignment and Matching. PhD Thesis, New York University (2016)
10. Mitzenmacher, M., Owen, S.: Estimating resemblance of MIDI documents. In: Proceedings of the 3rd International Workshop on Algorithm Engineering and Experimentation (ALENEX), Washington, DC, USA, pp. 78–90 (2001). https://doi.org/10.1007/3-540-44808-X_6

11. Schierle, C.: Visual MIDI Data Comparison. Master's Thesis, University of Stuttgart, Germany (2020). https://elib.uni-stuttgart.de/server/api/core/bitstreams/be2262f7-1b86-4b3e-b91d-6563d47c1768/content. Accessed 10 Nov 2024
12. Zolotov, A.V., Chernyshov, M.K.: On identification of MIDI music using fuzzy matching methods. In: Proceedings of the XII International Scientific-Practical Conference "Informatics: Problems, Methods, Technologies", pp. 149–150 (2012) (in Russian)
13. Lependin, A., Ladygin, P., Karev, V., Mansurov, A.: Fourier chromagrams for fingerprinting, verification and authentication of digital audio recordings. In: High-Performance Computing Systems and Technologies in Scientific Research, Automation of Control and Production (HPCST 2023). CCIS, vol. 1986, Springer, Cham. https://doi.org/10.1007/978-3-031-51057-1_20

Multithreaded Soft Processor Core with RISC-V Architecture

Aleksandr Kalachev(✉), Vladimir Pashnev, and Yuriy Matyuschenko

AltSU – Altai State University, Lenin Avenue 61, 656049 Barnaul, Russia
kalachev@phys.asu.ru

Abstract. This paper describes the development of an IP soft processor core based on the open and extensible RISC-V architecture implemented in Verilog. The design evolves from a multistage microarchitecture to a multithreaded core using shadow register sets, where each thread stores its architectural state and is dynamically mapped to general-purpose registers (x0–x31) and the program counter during execution. The multithreaded approach improves the efficiency of integrating the soft core into FPGA projects as a control unit by simplifying context switching and interrupt handling through dedicated shadow status registers. The architecture targets embedded systems with a single privilege level (machine mode), in which all threads operate within a unified administrative space, while protection and synchronization of shared data are handled at the software level. Following the RISC-V concept facilitates software development in high-level languages, although for small or highly specialized applications the use of virtual machines or domain-specific languages may be preferable. The implementation is reviewed for the minimal instruction set I+Zicsr, supporting integer operations and control and status registers, which are used to configure and control execution flow parameters for FPGA-based multithreaded control applications.

Keywords: RISC-V · IP-Cores · Soft-Processor · Multithreading · Hart · Prototyping

1 Introduction

Processors of the Xcore line from XMOS, a company that specializes in developing processors and microcontrollers for embedded systems, inspired the creation of a multithreaded version. Typically, XMOS solutions incorporate elements that usually demand components of a different class [1]. Previously, one might have needed a microcontroller to manage the structure, a digital signal processor for signal processing, and perhaps a programmable logic integrated circuit for interfacing with complex digital systems. However, XMOS processors can handle all these tasks within a single device through a unified software process. Key features include hardware multithreading, the capability to scale the number of cores or processors in the system, and adaptable software-configurable I/O ports. Initially, XMOS processors utilized proprietary RISC cores, and some models combined RISC and ARM cores. In 2023, the company introduced a processor variant featuring RISC-V cores (see Fig. 1).

V. Jordan et al. (Eds.): HPCST 2024, CCIS 2919, pp. 39–49, 2026.
https://doi.org/10.1007/978-3-032-20325-0_4

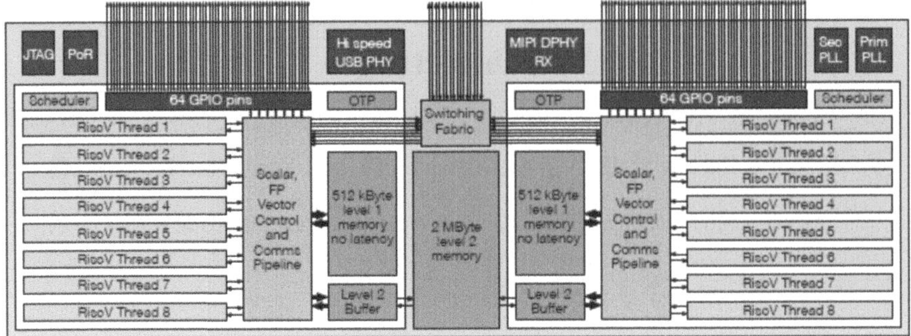

Fig. 1. Dual-core XMOS processor with eight-threaded RISC-V cores [2].

The second reason motivating the experiments is the FPGA software core development to avoid integrating an interrupt controller into the software processor. Developing a multithreaded microarchitecture for a software processor is important for FPGA-based systems for several reasons: -simplifying the handling of multiple asynchronous events through the use of multiple threads without the need for software context switching; -reducing the complexity of the interrupt controller; -simplifying the program code and decreasing its size.

One of the objectives of the work was to achieve compatibility between the multithreaded microarchitecture and the RISC-V architecture. The RISC-V architecture specification indicates support for multithreading in the form of hardware-supported threads. This feature was actually implemented in the microarchitecture under consideration. For each thread (hart), the thread state is saved in an array of shadow registers (main registers x0-x31 and the program counter) – each register set stores the architectural state of the thread. Shared resources include executive logic units (peripherals, ALUs), memory (programs and data). Additionally, all threads share a block of special-purpose registers (CSR) – status registers, timers, performance counters, etc.

2 Proposed Approach

2.1 Based RISC-V Core Architecture

Let's try to make changes to the multi-stroke processor version by dividing the steps of executing instructions in more detail and adding support for multithreading through "shadow" copies of the main architectural registers. In RISC-V terminology, the hardware-supported stream is designated as hart [3–6]. So far, the overall structure of the core will remain almost the same as shown in Fig. 2 [10].

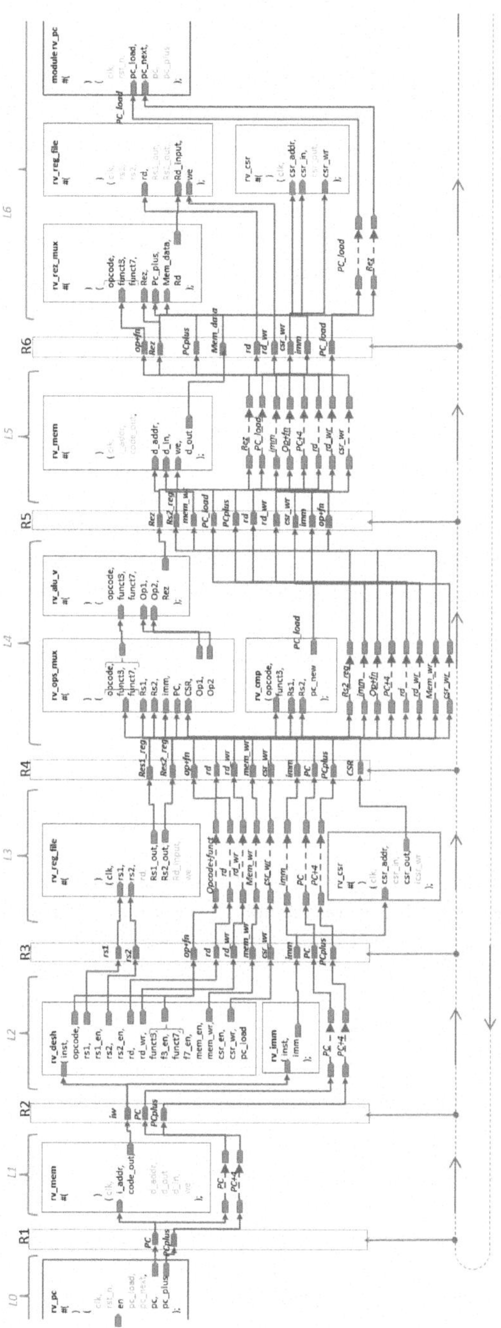

Fig. 2. Based RISC-V core multicycle microarchitecture

2.2 Multithreading Core Structure

The basic concept is that each hart stores the thread state in an array of shadow registers. In this case, each copy of the registers contains the architectural state of the stream. Logical blocks and memory are shared resources. It is also advisable to make a block of special function registers common to all harts (usually system timers, per-formance counters, etc.). In this case, the CSR register block must have two separate address inputs – for reading data and for writing to it (since it remains common to all harts, and read and write operations are performed at different stages).

Similarly, the file register must also be able to write and read data from registers associated with different harts. In fact, we define an array or, if you prefer, a stack of file registers. To control the selection of hart data, we will add a group of registers – hart-reg, combined into a ruler. The output signals will be the hart number and its activity bit. On reset, the registers are initialized to zeros, and the hart activity bits are reset to zero.

During operation, the ruler must receive a sequence of harts with activity codes. The sequence itself can be stored, for example, in a separate small memory – let's call it a hart_table. The hart_table table is addressed by a cyclic counter. For the cur-rent example, let's create a simple 64-bit "system timer" (although any suitable bit size is possible; here 64 bits are selected by analogy with traditional timer counters in RISC-V architecture implementations).

The low-order digits of the counter will ad-dress the memory of the hart table. The counter itself can be used, for example, as a clock cycle counter from the moment the system is started. The hart activity bits will also be needed to control writes to processor core modules such as the program counter, file register, and memory (if hart is inactive, no writes are performed). In addition, the hart feedback signal is applied to the input of the ALU "pipeline reset" (for this implementation, the ALU version of the processor core with a pipeline will be used).

The levels are "separated" by intermediate registers (ordinary storage registers) that capture data (registers R0–R6). Some signals simply "pass" in stages (directly between registers), if they are not involved in the current stage, but are required in subsequent ones – everything is the same as in the simple multi-stroke version. Description of the Hart table memory: the current version has a memory for 8 entries, the output depth is 4 bits, the highest bit is the hart activity bit, and the three lowest bits are the hart number. The system timer is a 64-bit incremental counter. Microarchitectural blocks of the multithreaded processor variant (see Fig. 3) [10]:

- rv_pc is a program counter (an array of several registers);
- rv_mem is a block of memory (software and operational);
- rv_desh is an instruction decoder (instruction words) with a "pipeline reset" input;
- rv_imm is a direct value generator from an instruction word;
- rv_reg_file is a register file (an array of several file registers);
- rv_ops_mux is an operand switch for ALU;
- rv_cmp is a transition resolution signal generator; rv_alu_v is ALU;
- rv_rez_mux is the switchboard of results;
- rv_csr – a block of special-purpose registers;
- rv_r_reg – registers for storing the results of intermediate stages;
- rv_hart_reg – registers for storing the numbers and activity status of the harts;

- rv_timer is the system timer;
- rv_hart_table is the storage memory for the list and status of harts.

The proposed version of the multithreaded RISC-V microarchitecture makes it possible for several independent or interacting programs to work within the same hardware core.

In this version, all program threads use shared memory, as the hardware does not include any means to protect data from simultaneous modification and writing. Therefore, ensuring the security of data and proper access to it becomes the task of the programmer. To comply with the RISC-V architecture, it is necessary to integrate the microarchitectural blocks of the system timer and the hart table into special-purpose register blocks (CSRs).

They can be changed programmatically using standard tools, which will ensure compliance with the requirements of the canonical architecture. In the future, it is planned to develop harts management mechanisms, both event–based (based on the principle of interrupts) and software-based (stopping or starting harts by the user or other harts).

To control sets of shadow registers, a group of registers is used – hart-reg, combined into a ruler. These registers output the hart number and its activity bit. On reset, the registers are initialized to zeros, and the hart activity bits are reset to zero. During operation, a sequence of harts with activity codes is fed into the hart-reg register string.

The thread sequence itself is stored in a separate small memory block, the hart_table. This table is accessed by a cyclic counter, the role of which is performed by the lower digits of the system timer.

The hart activity bits are used to control writing to processor core modules such as the program counter, file register, and memory (if hart is inactive, no writing is performed). The list of threads is stored in the rv_hart_table module, which contains the list and status of harts.

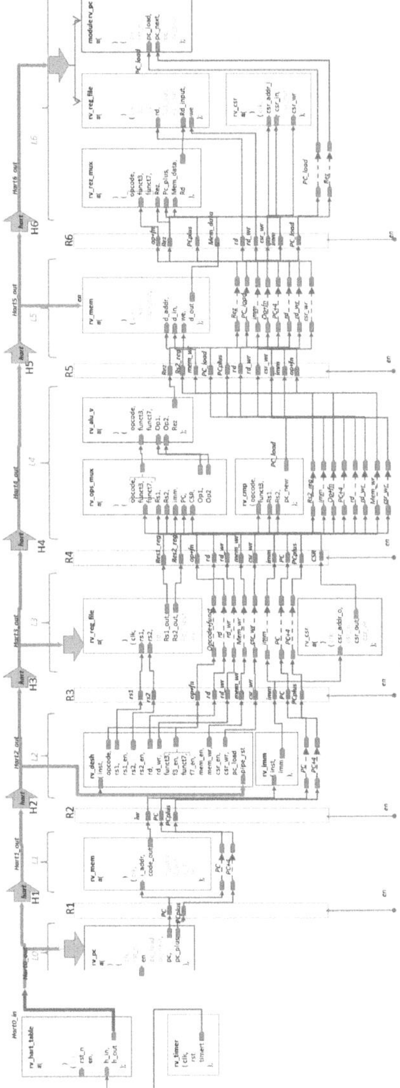

Fig. 3. Multithreading core structure.

2.3 Hardware Flow Management in the RISC-V Architecture

The RISC-V architecture specification does not include hardware flow control mechanisms: this function is assigned to software – application or system (the choice depends on the type and purpose of the processor). Within the framework of the architecture, only some special-purpose registers are fixed. These registers are: - they correspond to the access level to the CPU; - on systems with multi-level privilege, they are mapped to addresses accessible to unprivileged code.

- mhartid (read-only)– Hart ID Register – identifier of the hardware thread that executes the current code (for single–threaded processors mhartid = 0);
- mstatus (read/write) – Machine Status Register - the register contains the current state and controls the current state of the hardware thread.

It is important to note that the registers under consideration lack means of directly controlling the state of the flow (start/stop), especially for microcontroller–class cores. Thread interaction mechanisms are not standardized and implementation-dependent; they are usually coordinated by operating systems.

- through shared memory (often, to avoid data/variable races, it is probably better to use commands from the A-expansion of the command system, with its support), otherwise control is transferred to the software level;
- through interruptions/exceptions;
- using special-purpose registers (CSR registers) – this option has not been found in specification, but, formally, it is not prohibited in microarchitectural implementations.

Flow transition options between active and passive states. Programmatic (synchronous) events:

- independently (the stream has the ability to put itself in the "inactive" state - for example, by changing a variable, a bit in a dedicated CSR register);
- at the command of another thread (similarly, either through a common variable, or through a bit of the CSR register).

Asynchronous events are a reaction to:

- events of peripheral devices – timers, counters, transceivers (UART, SPI, I2C, I2S, etc.);
- events of data exchange between threads (read/write).

The solution to the problem involves the implementation of an interrupt controller with a strict allocation of responsibility: each event is processed by a dedicated hardware thread according to predefined conventions.

Key attributes of the stream:

- activity flag;
- program counter (defines the executable code).

There are two ways to modify counters:

- through the address space;
- -using CSR registers.

Transferring control to the CSR level is the most promising approach. In architectures with support for privileged commands, access to the CSR is controlled by the OS or system monitors. Using "remote" addresses (starting from 0x3800) avoids collisions with existing command system extensions. For example.

- 0x3800 – 0x381F - Hart_table - The table of threads in "tight rotation" – those sets of shadow registers, the execution of which is allowed or planned in the course of work – sets the order of execution of threads. The current stream is also displayed in

the current stream register (hartid) (the lower part), the stream activity bit is displayed in the stream status register (hart_state);
- 0x3820 – 0x382F - PC_table - Table of values of program flow counters in "tight rotation";
- 0x3840 – 0x384F - Sensivity_list - A set of bit fields showing how/by what means a stream can change its bit of activity.

The word bits in the Sensiviti_list fields encode possible conditions for the stream to change its state:

- the thread manages its state by itself;
- the flow is sensitive to the status and signals of system and peripheral devices - timers, interfaces;
- the stream reacts to the levels of the I/O lines.

3 Experiments and Results

3.1 Multithreading Core Testing – Simple Program

A small test program - the program contains several short infinite loops working with registers x1, x2. Register x1 is initiated by a certain value, register x2 will be used as a communication register. With a simple linear execution, the program will loop on the l1 label. Listing of the program is presented in Fig. 4.

For execution on a multithreaded processor, an array of program counters is initiated by the address values of the beginning of each cycle – in this example, the ad-dress:

0x0, 0x14, 0x28, 0x3C, 0x50, 0x64, 0x78, 0x8C.

The Hart table is an enumeration of the hart numbers from 0 to 7 in a 4-bit representation.

For example, if a heart with the number 0 should be active, then it will be represented in the table by the number 8 (4'b1000), if with the number 3, then 0xB (4'b1011) and so on (Fig. 5).

Address	Code	Basic	Line	Source
// for core				
#0	0x000010b7	**l1:** lui x1,1	1	lui x1,0x1
0x00400000	0x00c0d093	srli x1,x1,12	2	srli
0x00400004	0x00108093	addi x1,x1,1	4	x1,x1,12
0x00400008	0xffdff16f	jal x2,0xfffffffc	5	addi x1, x1,
0x0040000c	0x000990b7	lui x1,0x00000099	6	jal x2,l1
0x00400010				lui x1,0x99
// for core	0x000020b7	**l2:** lui x1,2	8	
#1	0x00c0d093	srli x1,x1,12	9	lui x1,0x2
0x00400014	0x00208093	addi x1,x1,2	11	srli
0x00400018	0xffdff16f	jal x2,0xfffffffc	12	x1,x1,12
0x0040001c	0x000990b7	lui x1,0x00000099	13	addi x1, x1,
0x00400020				2
0x00400024	0x000030b7	**l3:** lui x1,3	15	jal x2,l2
// for core	0x00c0d093	srli x1,x1,12	16	lui x1,0x99
#2	0x00308093	addi x1,x1,3	18	
0x00400028	0xffdff16f	jal x2,0xfffffffc	19	lui x1,0x3
0x0040002c	0x000990b7	lui x1,0x00000099	20	srli
0x00400030				x1,x1,12
0x00400034	0x000040b7	**l4:** lui x1,4	22	addi x1, x1,
0x00400038	0x00c0d093	srli x1,x1,12	23	3
// for core	0x00408093	addi x1,x1,4	25	jal x2,l3
#3	0xffdff16f	jal x2,0xfffffffc	26	lui x1,0x99
0x0040003c	0x000990b7	lui x1,0x00000099	27	
0x00400040				lui x1,0x4
0x00400044	0x000050b7	**l5:** lui x1,5	29	srli
0x00400048	0x00c0d093	srli x1,x1,12	30	x1,x1,12
0x0040004c	0x00508093	addi x1,x1,5	32	addi x1, x1,
//for core #4	0xffdff16f	jal x2,0xfffffffc	33	4
0x00400050	0x000990b7	lui x1,0x00000099	34	jal x2,l4
0x00400054				lui x1,0x99
0x00400058	0x000060b7	**l6:** lui x1,6	36	
0x0040005c	0x00c0d093	srli x1,x1,12	37	lui x1,0x5
0x00400060	0x00608093	addi x1,x1,6	39	srli
// for core	0xffdff16f	jal x2,0xfffffffc	40	x1,x1,12
#5	0x000990b7	lui x1,0x00000099	41	addi x1, x1,
0x00400064				jal x2,l5
0x00400068	0x000070b7	**l7:** lui x1,7	43	lui x1,0x99
0x0040006c	0x00c0d093	srli x1,x1,12	44	
0x00400070	0x00708093	addi x1,x1,7	46	lui x1,0x6
0x00400074	0xffdff16f	jal x2,0xfffffffc	47	srli
// for core	0x000990b7	lui x1,0x00000099	48	x1,x1,12
#6				addi x1, x1,
0x00400078	0x000080b7	**l8:** lui x1,8	50	6
0x0040007c	0x00c0d093	srli x1,x1,12	51	jal x2,l6
0x00400080	0x00808093	addi x1,x1,8	53	lui x1,0x99
0x00400084	0xffdff16f	jal x2,0xfffffffc	54	
0x00400088	0x000990b7	lui x1,0x00000099	55	lui x1,0x7
// for core				srli
#7				x1,x1,12
0x0040008c				addi x1, x1,
0x00400090				jal x2,l7
0x00400094				lui x1,0x99
0x00400098				
0x0040009c				lui x1,0x8
				srli
				x1,x1,12
				addi x1, x1,
				jal x2,l8
				lui x1,0x99

Fig. 4. Distribution of starting addresses and activity of harts – the 0th and 1st are active.

PC.txt	Hart_table.txt
@00	@00
0	8
14	9
28	2
3c	3
50	4
64	5
78	6
8c	7

Fig. 5. Distribution of starting addresses and activity of harts – the 0th and 1st are active.

3.2 Multithreading Core Testing - Simulation

Figure 6 shows time diagrams of code execution.

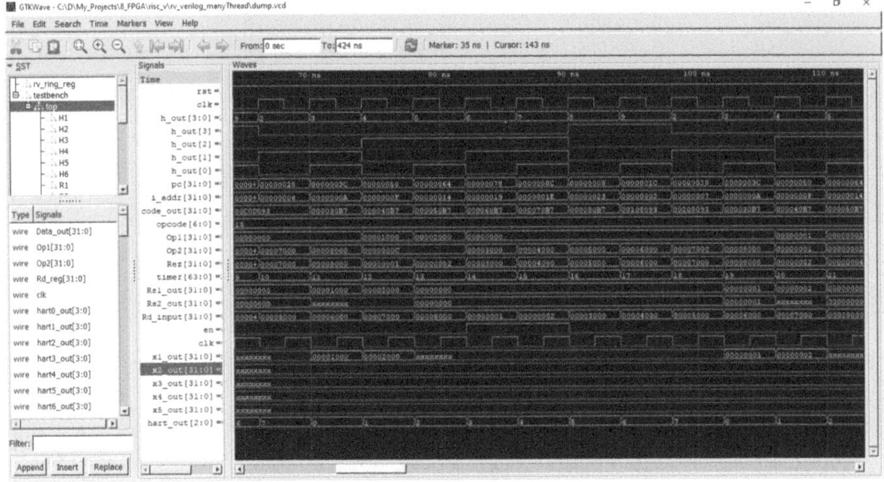

Fig. 6. Time diagrams of a multithreaded processor with two active harts

It shows the values:

- reset signals and clock;
- heart numbers (with the activity bit);
- the value of the program counter and the instruction word selected from memory;
- for ALU - command opcode, operands and result;
- for the file register, the current outputs and input are with write permission;
- test outputs of the first 4 registers of the file register, indicating which card they belong to.

For the X1 register of the 0th active hart, its loading with the value 0x1000 is visible first, then its shift to 0x1, similarly, for the 1st active hart, its loading with the value 0x2000 is visible, then its shift to 0x2.

4 Conclusion

The developed multithreaded RISC-V microarchitecture supports parallel execution of independent or coordinated software flows within a single hardware core. At the same time: shared memory is used without hardware protection mechanisms against competitive access.; ensuring data integrity and correctness of write/read operations is delegated to the software layer.

Compliance with the RISC-V architecture is achieved by: integration of system timer blocks and the hart table into special purpose registers (CSRs); the possibility of software modification of these blocks using standard tools, which ensures formal compliance with canonical requirements.

Further development directions include creation of harts management mechanisms, including event-based management (similar to interrupt handling) and software control (stop/start commands initiated by the user or other harts).

Characteristic features of the implementation: - equality of all streams and their functioning in a single address space; - shifting responsibility for data protection and access control to shared variables to the software layer; - support for a minimum set of I + Zicsr instructions, which does not conflict with current RISC-V standards and extensions.

References

1. XMOS: Multicore microcontrollers of the XCORE family. https://cyberleninka.ru/article/n/multiyadernye-mikrokontrollery-semeystva-xcore-ot-xmos. Accessed 10 Nov 2024
2. XMOS: Using RISC-V to define SoCs in software. https://www.xmos.com/using-risc-v-to-define-socs-in-software/. Accessed 10 Nov 2024
3. Waterman, A., Asanović, K.: The RISC-V Instruction Set Manual. Volume I: User-Level ISA, Version 2.2
4. Computer architecture: What is a hardware thread in RISC-V? Electrical Engineering Stack Exchange. https://electronics.stackexchange.com/questions/580645/what-is-a-hardware-thread-in-risc-v. Accessed 10 Nov 2024
5. Trusted Hart for Mobile RISC-V Security. IEEE Computer Society Proceedings. https://www.computer.org/csdl/proceedings-article/trustcom/2022/942500b587/1LFM9tik1IQ/. Accessed 10 Nov 2024
6. SHAKTI Development Team: RISC-V Assembly Language Programmer Manual, Part I (RISC-V Green Card). https://www.cl.cam.ac.uk/teaching/1617/ECAD+Arch/files/docs/RISCVGreenCardv8-20151013.pdf. Accessed 10 Nov 2024
7. Smith, S.: Programming in the RISC-V Assembly Language. Transl. by A.V. Logunov, ed. by A.Yu. Romanov. DMK Press, Moscow, 276 p.
8. Vivek SGT: Adding custom instructions compilation support to RISC-V toolchain. https://medium.com/@viveksgt/adding-custom-instructions-compilation-support-to-riscv-toolchain-78ce1b6efcf4. Accessed 10 Nov 2024
9. RISC-V for FPGA architecture, microarchitectural implementations. https://riscv-alliance.ru/material/risc-v-dlya-fpga-arhitektura-mikroarhitekturnye-realizaczii/. Accessed 10 Nov 2024
10. Creating RISC-V cores. https://riscv-alliance.ru/material/sozdanie-risc-v-yader/. Accessed 10 Nov 2024

Information Technologies and Computer Simulation of Physical Phenomena

Study of Structural-Phase Transformations in Matrix Homogeneous Compositions of Al@Ni Core-Shell Nanoparticles by MD-Simulation of SHS

Vladimir Jordan[1,2](✉) and Igor Shmakov[1]

[1] Altai State University, 61 Lenin Avenue, Barnaul 656049, Russia
jordan@phys.asu.ru

[2] Khristianovich Institute of Theoretical and Applied Mechanics, SB RAS, 4/1 Institutskaya Street, Novosibirsk 630090, Russia

Abstract. This paper analyzes the results of molecular dynamics simulation of SH-synthesis process of intermetallic compounds in matrix homogeneous compositions consisting of identical spherical Al@Ni nanoparticles with a core-shell structure in two variants: non-contact and contacting Al@Ni nanoparticles. The spherical core of Al@Ni nanoparticle contains a crystal lattice of Al atoms, in the spherical shell - a crystal lattice of Ni atoms. Simulation of SHS in a matrix structure with Al@Ni nanoparticles in contact with each other revealed a structural phase transition: the process of rapid crystallization of the Ni_3Al phase from the melt and recrystallization of the NiAl phase into the Ni_3Al phase. For computer simulation of SHS, the LAMMPS software package in a parallel version and the EAM potential of interatomic interaction were used. For visualization of the simulation results, the OVITO package and the authors' own programs were used.

Keywords: SH-synthesis · Molecular dynamics method · Crystal lattice · Unit cell · LAMMPS and OVITO software packages · Parallel computing

1 Introduction

In the last two decades, the development of modern computing technologies and computing systems has made it possible to perform high-performance calculations in the field of modern materials science, which is called computational materials science.

Computational materials science and engineering uses computer science, simulation techniques, and fundamentals of physical and chemical theory to understand how materials are created. The main goal of this branch of science includes the discovery of new materials, determination of the behavior and mechanisms of formation of materials, explanation of experiments and research of theories of creation of materials and their physical properties, as well as their functional characteristics. As an increasingly important subfield of materials science, computational materials science has similar specialties to computational chemistry and computational biology.

V. Jordan et al. (Eds.): HPCST 2024, CCIS 2919, pp. 53–68, 2026.
https://doi.org/10.1007/978-3-032-20325-0_5

The creation of functional materials with the required structure and necessary physical properties, for example, metal or metal-ceramic coatings on technical products, can be implemented quite effectively using the method of "self-propagating high-temperature synthesis" (SH-synthesis or SHS). The SHS-process occurs in an autowave mode in a fairly short time in the form of a chemical reaction of the initial reagents with significant exothermic heat release in the combustion front, self-propagating throughout the mixture of reagents. Initially, an external source of thermal energy (for example, using an electric coil) initiates the SHS-process for a short time in the so-called "ignition" zone, and subsequently the SHS-reaction is maintained due to the transfer of heat to the next layer of the reagent mixture. During the propagation of the SHS-combustion wave and after its completion in the "afterburning" zone, the process of formation of reaction products in the condensed phase continues for some time (for example, for Ni-Al and Ti-Al systems, various intermetallic compounds are formed). The composition and quantitative content of the resulting intermetallic phases are influenced by various factors, such as the stoichiometry and dispersion of the initial components, the initial temperature of the mixture heating and the ignition temperature of the SHS, etc.

This paper analyzes the results of molecular dynamics simulation of SH-synthesis process of intermetallic compounds in matrix homogeneous compositions consisting of identical spherical Al@Ni nanoparticles with a core-shell structure in two variants: non-contact and contacting Al@Ni nanoparticles (see Fig. 1). In the spherical core of each nanoparticle are Al atoms, which form a crystal lattice. In the spherical shell of each nanoparticle, Ni atoms also form a crystal lattice.

The article is devoted to the study of the kinetic features of SHS and structural-phase transformations occurring in homogeneous matrix compositions consisting of identical spherical Al@Ni nanoparticles (in two composition variants: non-contacting and contacting nanoparticles). The study is carried out using the method of computer molecular dynamics simulation using the LAMMPS [1] and OVITO [2] software packages, taking into account parallel computing, as well as the authors' own software developments [3, 4].

2 Methodology and Stages of MD-Simulation of SHS

The LAMMPS software package is based on the molecular dynamics method, and its compiled version of parallel computing corresponds to the SM-MIMD architecture using the MPI parallel programming interface, i.e. a mechanism for transmitting messages between computing nodes with distributed memory is used [5]. Under the control of the OpenMP interface, calculations are performed on each computing node with shared memory. This approach allows additionally accelerating the calculations in some cases. In addition, the EAM-potential (embedded atom model – EAM, version from 2009) was chosen as the interatomic interaction potential [6].

Figure 1 shows two variants of model matrix nanocrystalline structures with a uniform composition of Al@Ni nanoparticles (the first structure is made up of nanoparticles that are not in contact with each other, the second is made up of nanoparticles that are in contact), for which the results of a molecular dynamics study of structural changes and phase transformations were obtained [5].

Both matrix structures are volumetric (three-dimensional), and their schematic images of vertical flat sections are shown in Fig. 1. In each matrix structure, identical nanoparticles form a three-dimensional cubic structure of filling the volume with nanoparticles (3 × 3 × 3) in the amount of 27 pieces. In reality, due to the use of "periodic boundary conditions" in the simulation process using the LAMMPS package, it can be assumed that the matrix structure samples are not limited in size (or have large sizes in three dimensions).

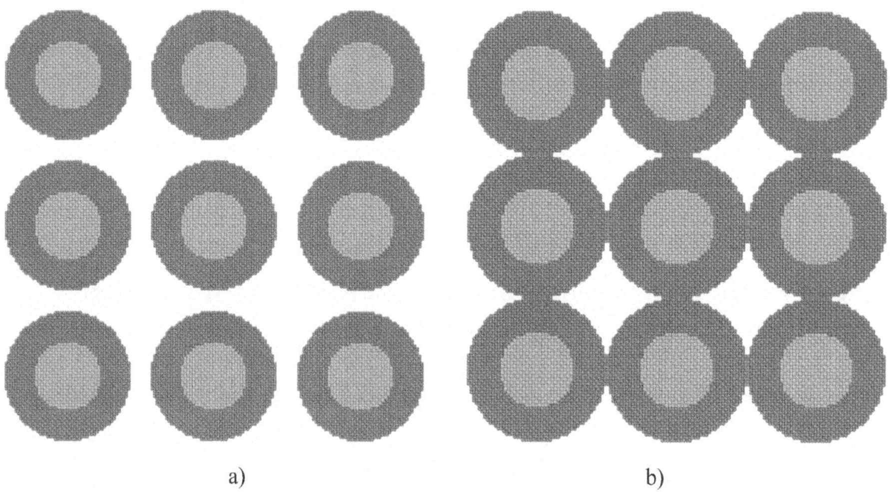

Fig. 1. Schematic images of vertical sections of two matrix nanocomposite structures (flat sections of the structures pass through the centers of the nanoparticles); Ni atoms are green; Al atoms are light gray [5]: a) structure of Al@Ni nanoparticles not in contact with each other; b) structure of Al@Ni nanoparticles in contact with each other. (Color figure olnine)

The study of the structure and phase formation during SH-synthesis using the example of a model matrix structure of filling the volume of the reaction mixture with composite nanoparticles (Fig. 1a) can be applied to study the features of structural-phase transformations during SH-synthesis in highly porous heterogeneous reaction compositions with nanostructured components (porosity over 50%, up to 80–90%), or, for example, in nanothermites [5, 7–10]. The model matrix structure with nanoparticles in contact with each other (Fig. 1b) is similar to a simple cubic lattice with a filling factor of about 0.524 (porosity 47.6%) and can be useful in studying SHS in heterogeneous compositions with an average degree of porosity (close to 50%) [5].

The crystal structure of the unit cells for Ni and Al is the same – it is a face-centered cubic lattice (fcc), but the lattice parameter values differ from each other. The lattice parameter for the lattice of Al atoms equal 0.405 nm [6]; for the lattice of Ni atoms it is 0.3524 nm [6].

For the two matrix structures, the simulation technique (and its stages) is the same. Each Al@Ni nanoparticle contains the same number of atoms $N = 33165$ ($N_{\mathrm{Al}} = 3511$, $N_{\mathrm{Ni}} = 29624$, stoichiometric ratio of atoms $N_{\mathrm{Al}}/N_{\mathrm{Ni}} = 0.1185$) [5]. The diameters of the nanoparticles (in the initial state) were the same in all cases (10 nm each).

In the first stage of the simulation (the "minimization" stage), the temperature is set to 600 K in both cases (Fig. 1a, b). The stage lasts 1.2 ps (or 600 iterations; one iteration corresponds to 0.002 ps of the computer experiment) [5].

In the second stage (the "relaxation" stage), fixed thermodynamic parameters of the canonical NPT-ensemble are set (pressure $P = 1$ bar, temperature $T = 600$ K and the number of N atoms for nanoparticles, respectively, 33165). In addition, during the simulation at this stage, periodic boundary conditions are maintained in all three dimensions for 0.4 ns. Periodic boundary conditions are maintained at all other stages of the simulation (for both structures) [5].

At the third stage of the simulation, the SHS-process "ignition" is initiated by means of rapid linear heating from 600 to 1200 K (for each structure) within 0.1 ns (within the NVT-ensemble, where V is the volume of the heating region). Then the SHS-process develops in the form of combustion wave propagation in the structure under study, while the conditions of the NVE-ensemble are observed (E is the total energy of the atoms of the system under study) [5].

3 Analysis of Kinetics and Structural-Phase Transformations in Matrix Structure with Non-Contacting Al@Ni Nanoparticles

For the structure (Fig. 1a) using Al@Ni nanoparticles that are not in contact with each other, the stoichiometry is $N_{\mathrm{Al}}/N_{\mathrm{Ni}} = 0.1185$ (slightly more than 8 Ni atoms per one Al atom). In the graph of Fig. 2, with an increase in time to 8 ns (liquid aluminum in the core of nanoparticles), a smooth increase in temperature to 1590 K is noticeable, and then the temperature remains unchanged up to the moment of 10.8 ns. Then, up to the moment of 11 ns, the temperature drops slightly (up to about 1580 K) and then slowly increases to approximately 1615 K (up to the moment of 15 ns). The temperature practically does not change in the next time interval (from 15 ns to 25.8 ns) [5].

To analyze the spatial (three-dimensional) distribution of temperature and density of a matter in the volume of nanoparticles, the authors previously developed and tested a method for "determining the 3D-density distribution depending on the 3D-distribution of temperature" based on a 3D-division of the nanoparticle volume into a cubic lattice consisting of domains (small cubes) [4]. In addition, based on the method of "determining the 3D-density distribution", a method of "recognizing the 3D-distribution of intermetallic phases by the domains-cubes of a cubic lattice" constructed for a nanoparticle inscribed in it was tested. For identical Al@Ni nanoparticles with a diameter of 10 nm, considered in this article, the cubic lattice of domains is a 3D-division in the form of a cube with dimensions of $10 \times 10 \times 10$, i.e. the division contains 1000 domains (a spherical Al@Ni nanoparticle is inscribed in a cube). Thus, in the eight corners of the cube, the domains (small cubes) are "empty" (in the figures given below in the text, such cases occur).

The temperature fluctuation in the domains of the Al@Ni nanoparticle (both positive and negative) relative to the temperature averaged over the entire volume of the Al@Ni nanoparticle turned out to be of the order of 80–90 K, and in individual domains in which the active formation of high-temperature intermetallic phases Ni_3Al and NiAl occurs, the fluctuation reached the order of 250–300 K. It should be noted that the

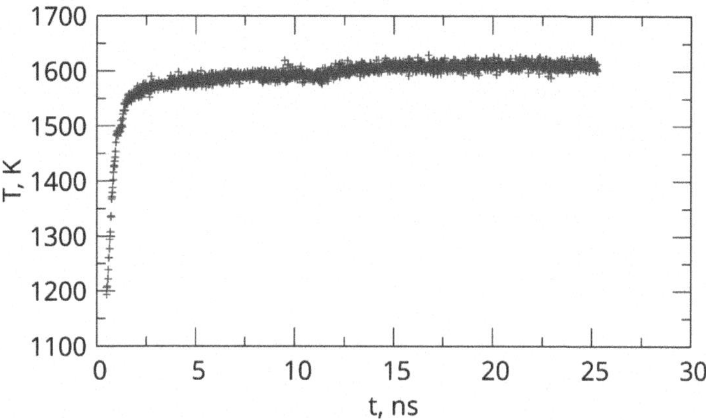

Fig. 2. Change of the average temperature at successive moments in time (the temperature is averaged over the entire volume of an Al@Ni nanoparticle and it is the same for all identical Al@Ni nanoparticles that are not in contact with each other)

average temperature is the same for all identical Al@Ni nanoparticles that are not in contact with each other, and the temperature fluctuations in the domains of Al@Ni nanoparticles are also the same.

A detailed analysis of structural-phase transformations in Al@Ni matrix nanoparticles (Fig. 1) was performed using the following methods built into the OVITO package [5]: 1) "the Common Neighbor Analysis" method (CNA) [2]; 2) the method of "angular and radial bonds between atoms" - Ackland-Jones analysis (hereinafter: AJ-analysis) [11]; 3) "the Polyhedral Template Matching" method (PTM) [12]. Regarding the effectiveness of methods for recognizing the crystal structures of intermetallic compounds, the following can be noted [5].

The CNA-analysis method uses information about interatomic distances (their threshold values) to classify the atomic structure. A disadvantage of the method is its sensitivity to thermal displacements and deformation, so pre-treatment such as quenching or time averaging of atomic arrangement is necessary [5]. Therefore, CNA-analysis is of little use for recognizing the types of crystalline unit cells in polycrystalline samples with temperatures close to melting points. The method is ineffective in recognizing atomic structures in amorphous samples [5].

The AJ-analysis method (taking into account the angular and radial distributions of bonds between atoms) is much more effective, since atoms that are incorrectly assigned or not assigned to certain types of crystal cells constitute a small number [5]. In addition, the AJ-analysis correctly reproduces the distant crystal structure for stable ideal crystals. The AJ-analysis method is quite effective in recognizing fcc-, bcc-, and hcp-structures.

In the process of PTM-analysis, it is possible to recognize a certain class of structural environment of atoms (environment topology) for each atom of the system, as a result of which it is possible to recognize the crystal structures of simple condensed phases in a local volume. By changing the threshold value of the RMSD-parameter (root mean square deviation), it is possible to minimize the measure of spatial deviation from the template of a particular ideal structure, i.e. to determine the quality of the correspondence of the topology of the atomic environment to a certain ideal crystal lattice (intermetallic phase). The structure type "Other (unrecognized type)" is assigned to atoms for which the RMSD-value exceeds the threshold value [5]. The threshold value is set to the most probable value - the peak of the RMSD-histogram. PTM-analysis is quite effective in recognizing structures in states close to the melting state, identifies common ordered structures of alloys, and local lattice orientation in polycrystalline samples [5, 12].

Figure 3 confirms the results of the comparison of the methods for analyzing the types of structures given above. Features in the kinetics of the SHS-process are demonstrated both in temperature kinetics (Fig. 2) and in structural-phase transformations (Figs. 3 and 4).

Fragments (a) and (b) of Fig. 3 (time points, respectively, 11.0 and 25.8 ns) confirm the fact that the CNA-analysis method in this case correctly identified fcc-structures (green color), bcc-structures (blue color) and hcp-structures (red color). In addition, the results of CNA-analysis correlate with the results of AJ- and PTM-analyses. Thus, all three methods of structural analysis confirmed the fact that the solid Ni-shell of the nanoparticle largely retains the crystalline fcc-structure (green color). A slight increase in fcc-structures and a slight decrease in bcc- and hcp-structures are noticeable in the AJ- and PTM-images in the liquid core up to the time point of 25.8 ns, starting from the time point of 11 ns (the characteristic moment of temperature dynamics) [5]. The increase in the number of fcc-structures in the core is associated with the formation of the Ni_3Al phase by crystallization from the melt, which at the beginning of the SHS-process in the core was liquid aluminum. Over time, crystalline nuclei of various intermetallic phases (including the Ni_3Al and NiAl phases) are formed in the liquid melt due to the reaction diffusion of Ni atoms from the solid-phase Ni-shell through the interface into the core of the nanoparticle. Additional confirmation of Fig. 3 are all fragments of Fig. 4 (the distribution of atoms and phase maps of the distribution of intermetallic phases in the "central" cross-section of the nanoparticle).

Analyzing the phase maps (fragments (c) and (d) of Fig. 4), single-phase domains 2 (Ni_3Al phase) are found in the region of the nanoparticle core, as well as other domains: two-phase domains (1; 2) – a mixture of nickel (fcc-structure) and Ni_3Al phases (also fcc-structure); one two-phase domain (2; 3) – a mixture of Ni_3Al and NiAl phases (bcc-structure) [5].

Returning to the fragments of Fig. 3, a decrease in the number of bcc-structures (NiAl phases – blue) and hcp-structures (red) is noticeable in the core, which can be explained to some extent by the recrystallization of the NiAl phase into the Ni_3Al phase of the NiAl phase into the Ni_3Al phase (CNA-analysis and PTM-analysis also confirm this fact). The recrystallization of the NiAl phase into the Ni_3Al phase is facilitated by the presence of a large excess of Ni in the initial Al@Ni nanoparticle (the stoichiometry of the Ni and Al components is slightly more than 8:1). The publication [13] confirms such a structural phase transition from the NiAl phase to the Ni_3Al phase in real experiments. In publications [14–18] the authors substantiate several mechanisms of crystallization of the Ni_3Al phase.

Namely, the mechanism of thermal explosion (high-temperature synthesis of Ni_3Al with a very high rate of exothermic heat release and sudden temperature increase) is discussed in the works [14, 15]. The works [16, 17] speak about the mechanism of

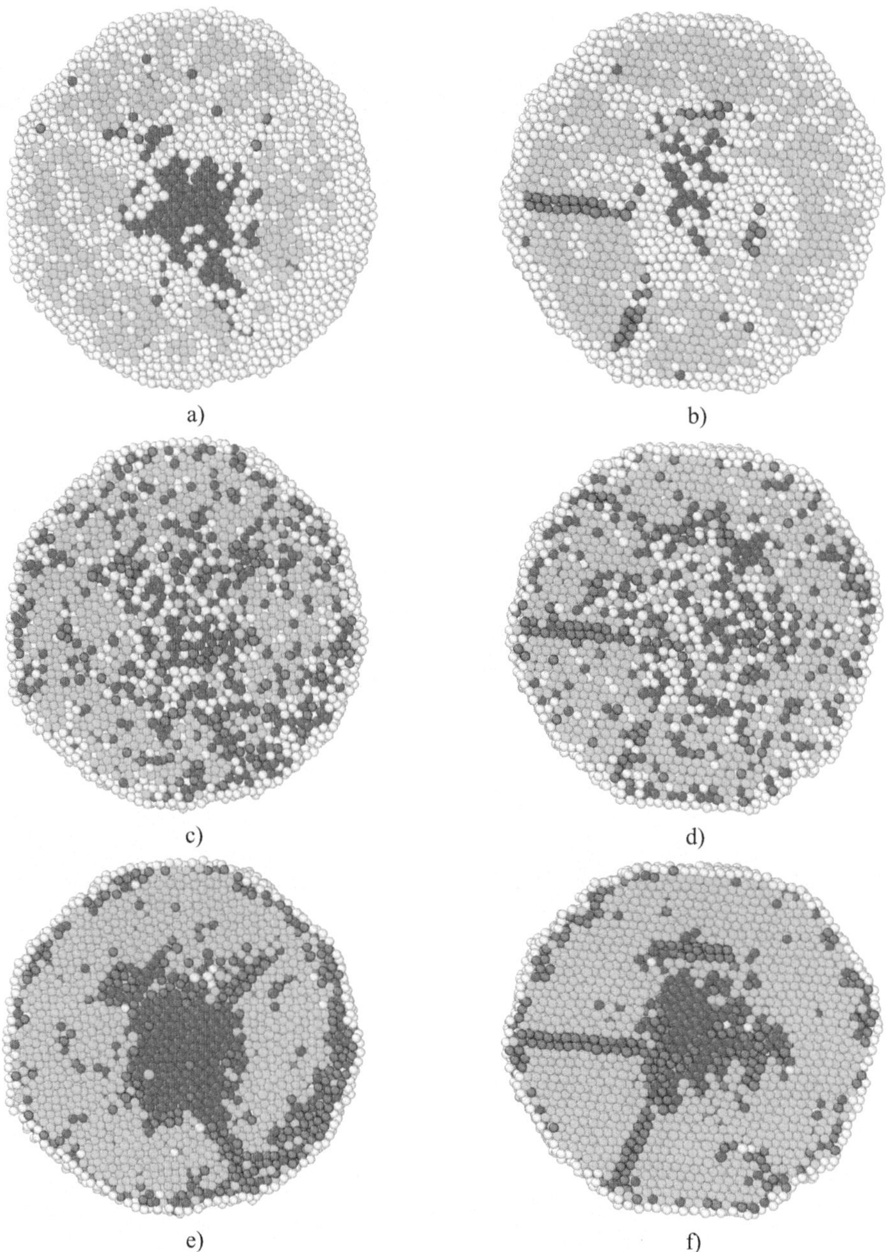

Fig. 3. Results of the analysis by three methods (CNA, AJ and PTM) of structural-phase transformations in Al@Ni nanoparticles not in contact with each other at the moments of time of 11 ns (fragments (a, c, e) and 25.8 ns (fragments (b, d, f). CNA-results: fragments (a) and (b); AJ-results: fragments (c) and (d); PTM-results: fragments (e) and (f). Color designations of the structures: bcc – blue, hcp – red, fcc – green, icosahedron – yellowish-brown, "unrecognized type" – light-gray shade [5]. (Color figure olnine)

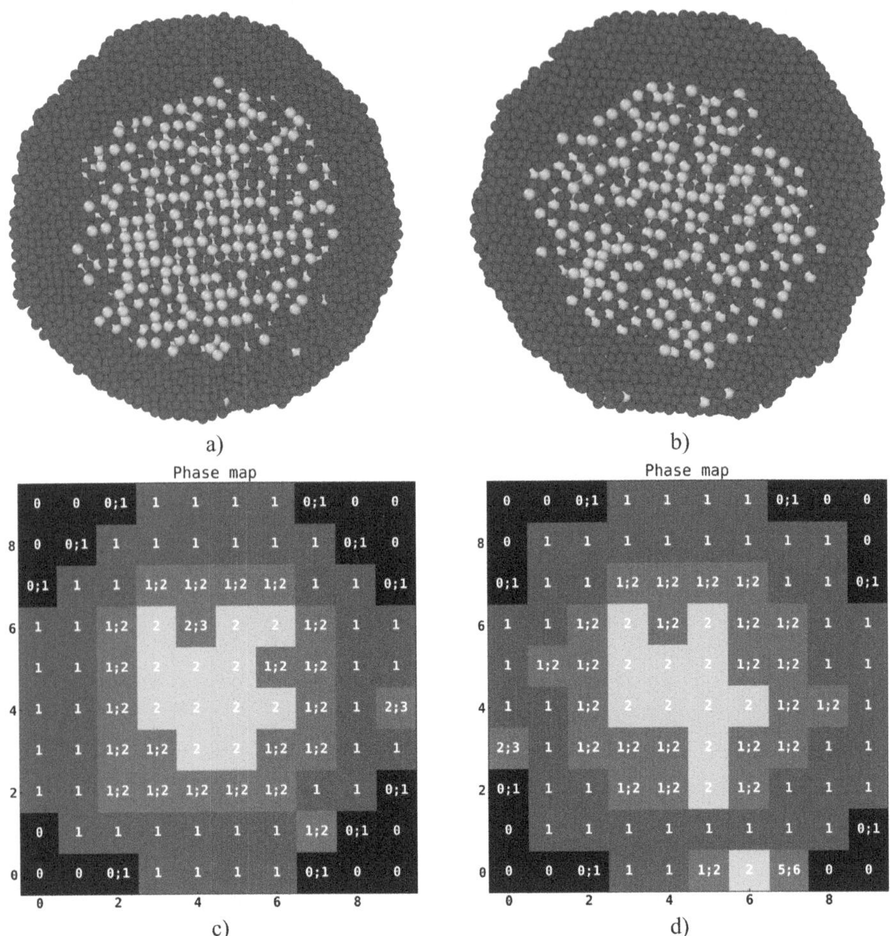

Fig. 4. Atomic distribution (Ni – blue; Al – gray): a) $t = 11$ ns; b) $t = 25.8$ ns. Phase maps for Al@Ni nanoparticles not in contact with each other: c) $t = 11$ ns; d) $t = 25.8$ ns. Explanation of phases on the maps: 1 – Ni, 2 – Ni_3Al, 3 – NiAl, 4 – Ni_2Al_3, 5 – $NiAl_3$, 6 – Al, 0 – empty domain; (1;2), (2;3) and (5;6) – domains with two-phase mixtures of the corresponding homogeneous phases; (0;1) – domain partially occupied by nickel [5].

Ni_3Al synthesis as a combination of two smooth processes occurring in the afterburning zone of a powder mixture of pure metals after the peak of maximum heat release. Crystallization from a cooling melt is the first process, and the diffusion interaction of intermediate products with the initial reagents is the second process. The process of SH-synthesis with phase transformations described in this paragraph 3 fits into the above-mentioned combined mechanism of a “smooth” nature. The mechanism of Ni_3Al synthesis, described in the following paragraph 4, is considered in publications [18, p. 41] and it occurs at a peritectic transformation temperature of 1396 °C and slightly higher (i.e., at 1669 K and slightly higher) [5].

To summarize, we note that the SHS process in a practical sense does not violate the original spherical shape of the Al@Ni nanoparticles, which are not in contact with each other (the case considered in this paragraph 3) [5].

4 Analysis of Kinetics and Structural-Phase Transformations in Matrix Structure with Al@Ni Nanoparticles in Contact with Each Other

For the structure (Fig. 1b) with Al@Ni nanoparticles in contact with each other, the stoichiometry is $N_{Al}/N_{Ni} = 0.1185$ (there are slightly more than 8 Ni atoms per one Al atom). In the graph of Fig. 5 for Al@Ni nanoparticles in contact with each other (liquid aluminum in the core), a smooth increase in temperature is noticeable with increasing time in the range up to 7.4 ns, and then (Fig. 5) a sharp jump is observed in temperature from 1645 to 1670 K [5].

Below is an analysis of the simulation results confirming the fact of the formation of the Ni_3Al phase at a peritectic transformation temperature of 1669 K, which is also confirmed in the work [18, p. 41]. In the domains of the Al@Ni nanoparticles, the temperature fluctuates (both positively and negatively) by about 100 K relative to the average temperature, and in individual domains the fluctuation reached about 300 K. In these domains, active formation of high-temperature intermetallic phases Ni_3Al and NiAl occurs (in two or three domains it reached about 400 K) [5]. It should be noted that the average temperature is the same for all identical Al@Ni nanoparticles in contact with each other, and the temperature fluctuations in the corresponding domains of identical Al@Ni nanoparticles included in the matrix structure are also the same (Fig. 1b).

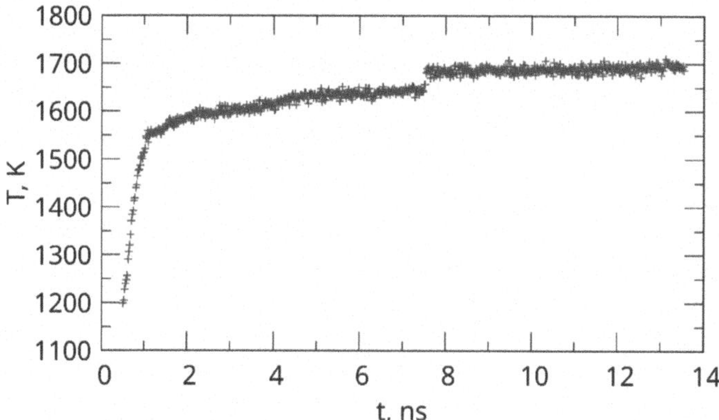

Fig. 5. Change in temperature averaged over the entire volume of an Al@Ni nanoparticle at successive moments in time (the average temperature is the same for all identical Al@Ni nanoparticles in contact with each other).

As noted at the end of paragraph 3, in the SHS-process, the nanoparticles in the structures (Fig. 1a) that are not in contact with each other mainly retain their original spherical surface shape. For the structure (Fig. 1b) with Al@Ni nanoparticles in contact with each other, their spherical shape changes significantly in the SHS-process. Figure 6 shows cross-sections of the matrix structure (at time $t = 13.5$ ns) by three mutually perpendicular planes (XY, XZ, YZ) passing through the centers of the nanoparticles.

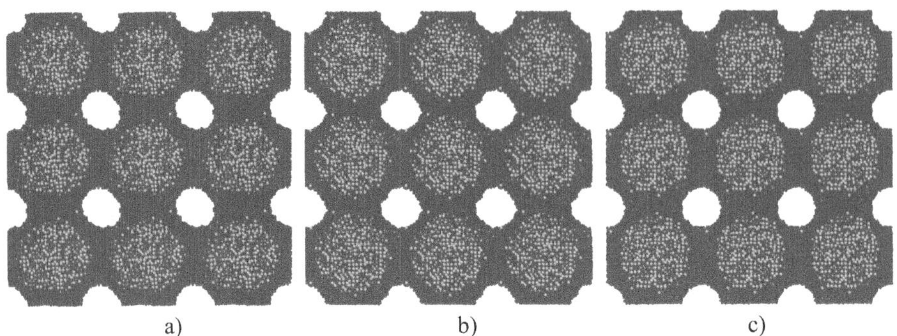

Fig. 6. Images of cross-sections of the matrix structure by three planes (XY, XZ, YZ) at time $t = 13.5$ ns for Al@Ni nanoparticles in contact with each other: fragment (a) corresponds to the cross-section by the XY-plane; (b) to the cross-section by the XZ-plane; (c) to the cross-section by the YZ-plane. Al atoms are marked in light gray, Ni atoms in blue. (Color figure online)

Nickel, located in the shell of Al@Ni nanoparticles in contact with each other, corresponds to a solid-phase state, since the average combustion temperature of SHS in nanoparticles does not exceed 1700 K and it is less than the melting temperature of nickel, equal to 1728 K. Therefore, the solid-phase Ni-shells of nanoparticles do not undergo a spreading process. However, closely located Ni atoms from the shells of neighboring nanoparticles (at the boundary of the joint contact) are attracted (come closer) due to the EAM-potential, which leads to an increase in the joint contact area of neighboring nanoparticles and to a change in their surface topology. But in three Cartesian planes (in three sections, see Fig. 6a–c) the symmetry of the form is preserved [5].

Thus, the structure (Fig. 1b) in three directions X, Y and Z, perpendicular to the planes, respectively, YZ, XZ and XY, has through "tubular" channels.

For the structure with Al@Ni nanoparticles in contact with each other (Fig. 1b), in comparison with the structure of non-contacting Al@Ni nanoparticles (Fig. 1a), differences are observed in the kinetics of the SHS-process, both in the temperature kinetics (a jump is noticeable – Fig. 5), and in the structural-phase transformations and changes in the shape of the nanoparticles (Figs. 7 and 8) [5].

As can be seen from Fig. 7, all three methods confirmed the preservation of the crystalline fcc-structure in the solid Ni-shell. During the temperature jump from 7.4 ns to 7.5 ns (Fig. 5), the CNA-analysis in the liquid core of the nanoparticle reflects an increase in the number of fcc-structures (the appearance of nuclei of the Ni_3Al phase; Fig. 7a, b) only to a certain extent, and practically does not detect other changes. The results of the AJ- and PTM-methods, correlating with each other, jointly confirm the intensive growth of the number of fcc-structures and a sharp decrease in the number of hcp-structures. In the core of the Al@Ni nanoparticle, PTM-analysis, compared to AJ-analysis, showed a slightly greater increase in the number of fcc-structures, which better corresponds to the phase map (Fig. 8d).

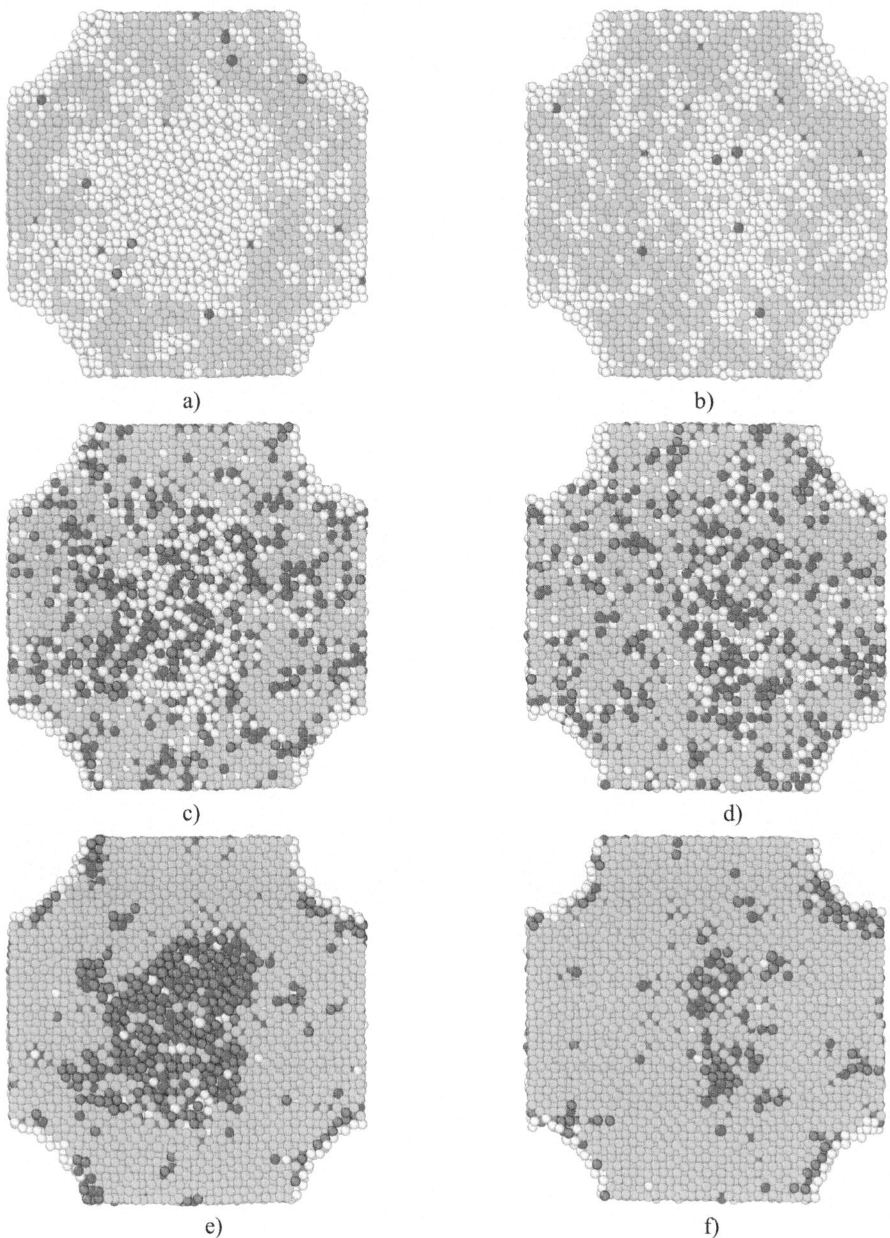

Fig. 7. Results of the analysis by three methods (CNA, AJ and PTM) of structural-phase transformations in Al@Ni nanoparticles in contact with each other at the moments of time of 7.4 ns (fragments (a, c, e) and 7.5 ns (fragments (b, d, f). CNA-results: fragments (a) and (b); AJ-results: fragments (c) and (d); PTM-results: fragments (e) and (f). Color designations of the structures: bcc – blue, hcp – red, fcc – green, icosahedron – yellowish-brown, "unrecognized type" – light-gray shade. (Color figure olnine)

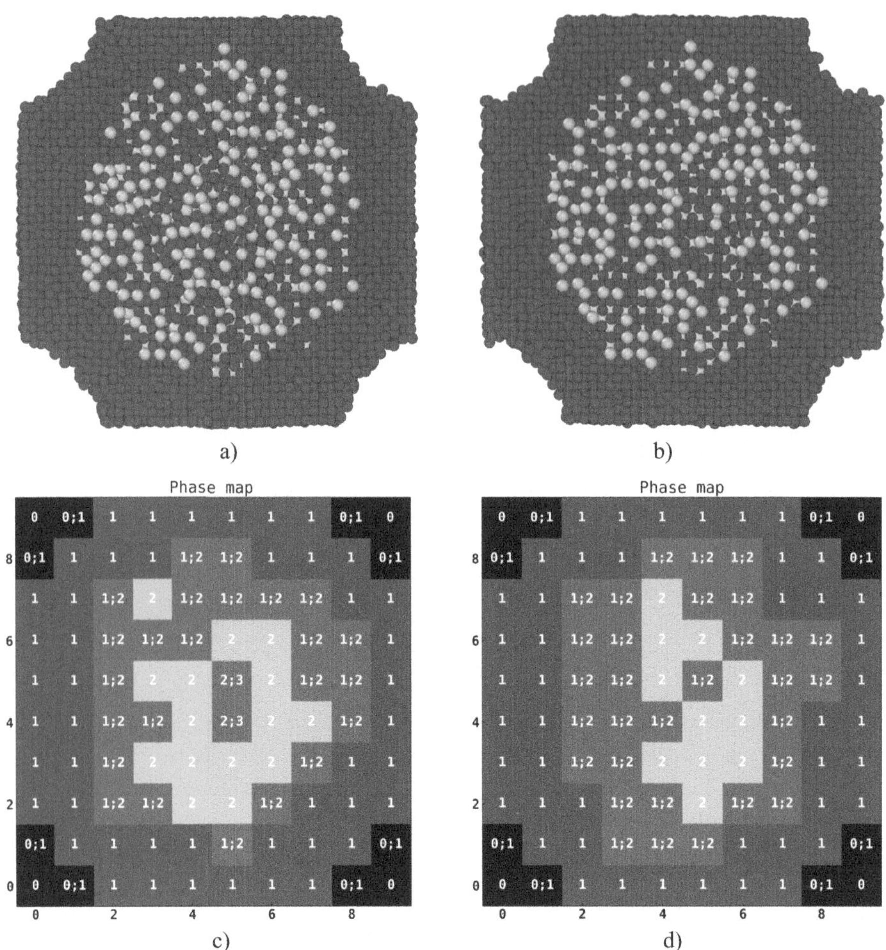

Fig. 8. Atomic distribution (Ni – blue; Al – gray): a) t = 7.4 ns; b) t = 7.5 ns. Phase maps for Al@Ni nanoparticles in contact with each other: c) t = 7.4 ns; d) t = 7.5 ns. Explanation of phases on the maps: 1 – Ni, 2 – Ni_3Al, 3 – NiAl, 4 – Ni_2Al_3, 5 – $NiAl_3$, 6 – Al, 0 – empty domain; (1;2), (2;3) – domains with two-phase mixtures of the corresponding homogeneous phases; (0;1) – domain partially occupied by nickel. (Color figure olnine)

By analyzing the changes in the distributions of Ni and Al atoms at times of 7.4 ns and 7.5 ns (fragments (a) and (b) of Fig. 8) in the core of the nanoparticle, one can talk about an increase in the degree of ordering of the structures in a very short time. This structural rearrangement in the core of a nanoparticle corresponds to the so-called "structural phase transition" [19, 20]. Namely, in the core of the nanoparticle, fragment (b) of Fig. 8 (in relation to fragment (a) of Fig. 8) more clearly demonstrates vertical and horizontal rows of Al atoms (gray color), alternating with Ni atoms (blue color) and lying with them in the same plane (on the same face), which confirms the increase in the number of fcc-structures by the time of 7.5 ns. By analyzing fragments (c) and (d) of the same Fig. 8 (phase maps), phases 2 (Ni_3Al – fcc-structure) and 3 (NiAl – bcc-structure) are detected, which correlate (according to their location on the phase maps) with the local fcc- and bcc-structures reflected in the corresponding images (fragments (a) and (b) of Fig. 8) [5].

Thus, the formation of the Ni_3Al phase in the nanoparticle core in a very short time (less than 0.1 ns) occurs in the mode of rapid crystallization from the melt and is accompanied by a sharp jump in temperature to the peritectic transformation temperature of 1669–1670 K (see Fig. 5). Confirmation of the recrystallization of the NiAl phase into the Ni_3Al phase is the disappearance of phase 3 (NiAl), present in fragment (c) of Fig. 8, and the appearance of phase 2 (Ni_3Al) in these places, according to fragment (d) of Fig. 8. The result of structural-phase transformations in the core of nanoparticles is the dominance of the Ni_3Al phase with the presence of excess Ni, which corresponds to the general stoichiometric ratio of the Ni and Al components of the original Al@Ni nanoparticle [5]. A sharp jump in temperature in a time of less than 0.1 ns characterizes a jump in the rate of heat release similar to the thermal explosion mode.

Below, in fragments (a) and (c) of Fig. 9, as an example, temperature maps are shown for the central section of a nanoparticle at times of 7.4 ns and 7.5 ns, in which individual domains with flashes of thermal explosion are clearly visible.

As an example, density maps corresponding to temperature maps (fragments (a) and (c) of Fig. 9) are shown in fragments (b) and (d) of Fig. 9 [5].

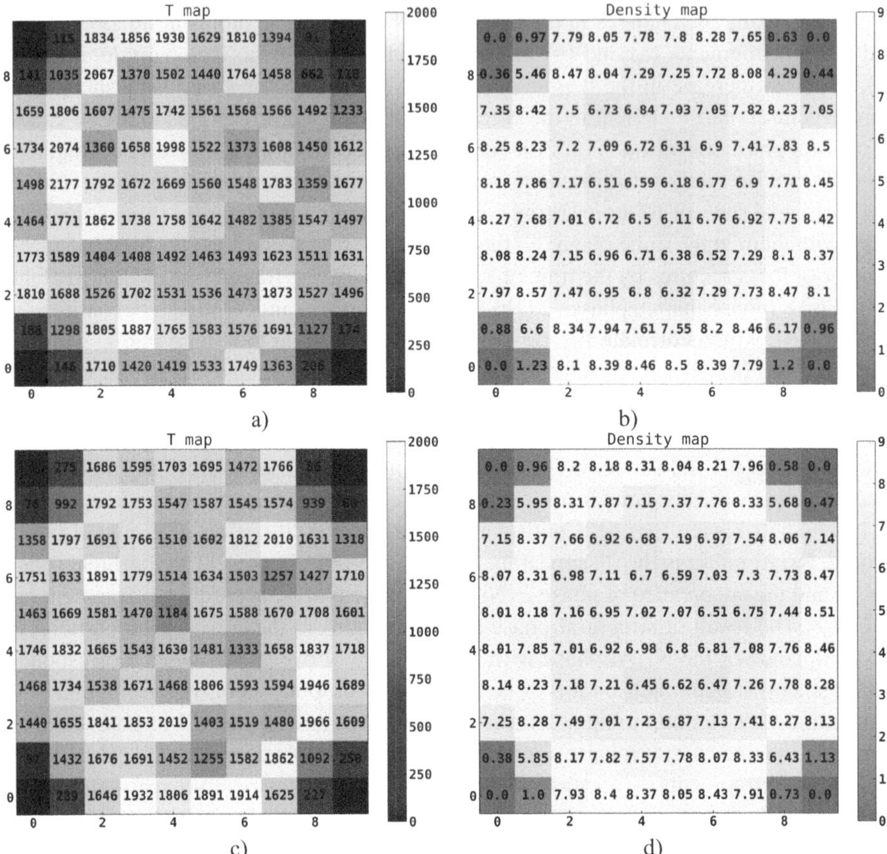

Fig. 9. Temperature distribution maps in domains of Al@Ni nanoparticles (in contact with each other): a) $t = 7.4$ ns; c) $t = 7.5$ ns. Density distribution maps in domains of Al@Ni nanoparticles (in contact with each other): b) $t = 7.4$ ns; d) $t = 7.5$ ns.

5 Conclusion

As significant results for two matrix structures (Fig. 1), recognition maps of intermetallic phases synthesized in nanoparticles at successive moments of time are presented. In the structure with Al@Ni nanoparticles in contact with each other, their spherical shape changes during the SHS-process. One of the significant differences between the results of this article and the results of work [21], devoted to phase formation in Al@Ni and Ni@Al nanoparticles in the "temperature annealing" mode (this mode differs from SHS in its strict controllability, rather than spontaneity), is as follows. In contrast to the publication [21], in this article, considering the variant of Al@Ni nanoparticles in contact with each other in a matrix structure (with a significant predominance of the number of Ni atoms over the number of Al atoms), the process of intensive reactive diffusion of Ni atoms from the shell into the liquid Al core at a certain moment initiates a structural phase transition in the nanoparticle core, consisting of rapid crystallization of the Ni_3Al

phase from the melt with additional recrystallization of the NiAl phase into the Ni_3Al phase. In addition, in the same short time, not exceeding 0.1 ns, there is a sharp jump in temperature and a jump in the rate of heat release (similar to a thermal explosion). The practical significance of the analyzed results is that they may be useful in the creation of porous nanostructured composite materials [5].

The use of the LAMMPS software package in the parallel computing version allowed us to achieve high efficiency of computational experiments. A cluster on 2 nodes was used as a computing platform. Characteristics of each node: 2 Intel Xeon E5-2690v3 processors, 64 GB RAM. Total: 48 cores, 98 threads, communication between nodes via InfiniBand Mellanox 40G/56G.

References

1. Plimpton, S.: Fast parallel algorithms for short-range molecular dynamics. J Comp. Phys. **117**, 1–19 (1995)
2. Stukowski, A.: Visualization and analysis of atomistic simulation data with OVITO – the open visualization tool. Model. Simulat. Mater. Sci. Eng. **18**, 015012 (2010)
3. Jordan, V.I., Shmakov, I.A: Computational procedure for the recognition of synthesized intermetallic interlayers at the interface in a Ni@Al "core-shell" nanoparticles. High-Perform. Comput. Syst. Technol. **5**(2), 42–52 (2021). (in Russian)
4. Jordan, V.I., Shmakov, I.A.: Method for Intermetallide spatial 3D-distribution recognition in the cubic Ni@Al "Core-Shell " nanoparticle based on computer MD-simulation of SHS. Commun. Comput. Inform. Sci. **1526**, 101–120 (2022). https://link.springer.com/chapter/10.1007/978-3-030-94141-3_9
5. Jordan, V.I., Shmakov, I.A.: molecular dynamics study of structural-phase transformations during SH synthesis of nickel aluminides in matrix structures with a homogeneous composition of nanoparticles of the core-shell composite structure. Sci. Notes Phys. Depart. Moscow Univ. **6**, 2460501 (2024). (in Russian)
6. Purja, P.G.P., Mishin, Y.: Development of an interatomic potential for the Ni-Al system. Phil. Mag. **89**(34–36), 3245–3267 (2009)
7. Rogachev, A.S., Mukasyan, A.S.: Combustion for materials synthesis: an introduction to structural macrokinetics, 400 p. Fizmatlit, Moscow (2012). (in Russian)
8. Politano, O., Baras, F.: Molecular dynamics simulations of self-propagating reactions in Ni-Al multilayer nanofoils. J. Alloy. Compd. **652**, 25 (2015). https://doi.org/10.1016/j.jallcom.2015.08.134
9. Turlo, V., Politano, O., Baras, F.: Microstructure evolution and self-propagating reactions in Ni-Al nanofoils: an atomic-scale description. J. Alloy. Compd. **708**, 989 (2017). https://doi.org/10.1016/j.jallcom.2017.03.051
10. Rogachev, A.S., et al.: Combustion in reactive multilayer Ni/Al nanofoils: experiments and molecular dynamic simulation. Combust. Flame **166**, 158169 (2016). https://doi.org/10.1016/j.combustflame.2016.014
11. Ackland, G.J., Jones, A.P.: Applications of local crystal structure measures in experiment and simulation. Phys. Rev. B. **73**(5), 054104 (2006). https://journals.aps.org/prb/abstract/10.1103/PhysRevB.73.054104
12. Larsen, P.M., Schmidt, S., Schiøtz, J.: Robust structural identification via polyhedral template matching. Modell. Simul. Mater. Sci. Eng. **24**(5), 055007 (2016). https://doi.org/10.1088/0965-0393/24/5/055007

13. Shmorgun, V.G., Bogdanov, A.I., Taube, A.O., Serov, A.G.: Transformation of chemical and phase composition of the Al–Ni and Al–Ni–Cr laminated coatings after high-temperature heating. Izvestiya vuzov. Poroshkovaya metallurgiya i funktsional'nye pokrytiya **1**, 51–59 (2016). https://doi.org/10.17073/1997-308X-2016-1-51-59. (in Russian)
14. Lapshin, OV., Ovcharenko, V.E.: Mathematical model of high-temperature synthesis of nickel aluminide Ni_3Al in the mode of thermal explosion of a powder mixture of pure elements. Phys. Combust. Explos. **32**(3), 68–76 (1996). (in Russian)
15. Lapshin, OV., Ovcharenko, V.E., Boyangin, E.N.: Thermokinetic and thermophysical parameters of high-temperature synthesis of nickel aluminide Ni_3Al in the mode of thermal explosion of a powder mixture of pure elements. Phys. Combust. Explos. **38**(4), 59–64 (2002). (in Russian)
16. Itin, V.I., Nayborodenko, Yu.S.: High-temperature synthesis of intermetallic compounds, 214 p. Tomsk University Press, Tomsk (1989). (in Russian)
17. Lapshin, OV., Boyangin, E.N., Ovcharenko, V.E.: Thermokinetic characteristics of the final stage of thermal explosion of powder mixture 3Ni+Al+TiC. Phys. Combust. Explos. **41**(1), 73–79 (2005). (in Russian)
18. Ibraeva, G.M.: Multilayer structure of intermetallic compounds of cobalt, nickel and titanium aluminides. Diss. for the degree of PhD in the specialty 6D071000 - Materials Science and Technology of New Materials, 91 p. (2019). Almaty. Republic of Kazakhstan
19. Myagkov, V.G., Bykova, L.E., Bondarenko, G.N., Bondarenko, G.V., Myagkov, F.V.: Solid-state reactions and order-disorder phase transition in thin films. J. Tech. Phys. **71**(6), 104–109 (2001). (in Russian)
20. Aksenov, V.L., Tropin, T.V.: Lectures on the theory of condensed matter. Textbook, 442 p. Phys. Faculty of M.V. Lomonosov Moscow State University Press, Moscow (2020). (in Russian)
21. Kart, S.O., Kart, H.H., Cagin, T.: Atomic-scale insights into structural and thermodynamic stability of spherical Al@Ni and Ni@Al core–shell nanoparticles. J. Nanopart. Res. **22**(140), 1–19 (2020). https://doi.org/10.1007/s11051-020-04862-2

Numerical Simulation of Thawing Processes in Frozen Soil Under Natural Effects and Anthropogenic Effects Associated with Well Operations

Sergey I. Markov[1,2](✉), Anastasia. Yu. Kutishcheva[1,2], Natalya B. Itkina[2,3], and Ella P. Shurina[1,2]

[1] Trofimuk Institute of Petroleum Geology and Geophysics, SB RAS, Koptug Avenue 3, 630090 Novosibirsk, Russia
www.sim91@list.ru

[2] Novosibirsk State Technical University, Karl Marx Avenue 20, 630073 Novosibirsk, Russia

[3] Institute of Computational Technologies, SB RAS, Academician M.A. Lavrentiev Avenue 6, 630090 Novosibirsk, Russia

Abstract. We present numerical procedure for mathematical simulation the seasonal change in temperature of frozen rocks during natural and constant temperature influence from a drilled well. We apply an effective model of the upper layer of frozen rocks to study the influence of seasonal changes in the temperature background during the annual period, taking into account the borehole surface temperature effect on the degradation process of permafrost rocks. These rocks are characterized by high intensity of moisture phase transitions at positive temperatures. Two mathematical models are used to simulate the above processes. The first model is based on the Stefan problem with explicit tracking of the phase transition front in a medium with inclusions having phase-changing properties. The second mathematical model is based on the heat conduction problem with phase transformations in enthalpy formulation. To reconcile the two mathematical models, we require continuity of the temperature and the normal component of the heat flux at the interface of the two domains. This approach allows us to take into account the natural multiscale nature of the object under study. For spatial discretization of mathematical models, the computational scheme of the discontinuous Galerkin method is used. The Runge-Kutta scheme is used for time discretization. Due to the block structure of the resulting system of algebraic equations, parallel realization of the problem solver is possible. To study the structural changes of frozen soil during thawing, the problem of numerical simulation of the thermoelastic deformation process is posed. The discretization of mathematical model is performed using the heterogeneous multiscale finite element method. We analyzed the results of computational experiments. We found that the borehole has a significant effect on the process of soil thawing during periods of sub-zero temperatures. In periods of positive day and night temperatures, phase transitions depend less on the borehole temperature. During the transition periods from positive to negative temperatures, two fronts of phase transitions are observed: from the day surface and from the deep-frozen ground layers. Thus, there are stress states that can lead to changes in soil structure in permafrost areas.

V. Jordan et al. (Eds.): HPCST 2024, CCIS 2919, pp. 69–84, 2026.
https://doi.org/10.1007/978-3-032-20325-0_6

Keywords: Frozen Soil · Thawing Process · Stress State · Numerical Simulation · Finite Element Method

1 Introduction

Construction and operation of engineering structures for oil and oil products extraction leads to significant changes in geocryological conditions in the Arctic zone. Progression of negative processes affecting the stability of engineering structures in the permafrost zone occurs even with a small change in the thermal balance of the environment. One of the methods of studying these effects is based on the application of numerical simulation of thermal processes with phase transitions in frozen soils.

In [1–4] thermal engineering calculations of the site with various technical structures are carried out, and permafrost thawing halos are studied. In [5] the permafrost along the Mohe-Datsin pipeline is monitored. This object is characterized by high temperature, high ice con tent, high sensitivity to external influences and very low thermal stability. Numerical simulation of temperature distribution around the pipes is performed. Ready-made software products are used: ANSYS, MATLAB, Abaqus, and others.

In [6] the processes of seasonal changes in the upper layers of permafrost soils are described. The heat conduction equation is considered, and an algorithm for its solution with computational optimization is developed. The proposed parallel algorithm is realized for a multicore processor using the OpenMP technology. The authors consider a three-dimensional model that allows describing the heat distribution in the upper layers of permafrost soils, taking into account the most significant climatic factors (seasonal changes in temperature and solar radiation intensity due to the geographical location of the field), physical factors (various thermal characteristics of heterogeneous soil varying in time), and the peculiarities of engineering construction of production wells and other types of technical systems.

In [7] the problem of stability of a pile foundation in permafrost regions is solved. A theoretical model of pile performance based on the principle of superposition and coordination dependence of pile and soil deformations is constructed. Finite-difference schemes and the MATLAB software package are used to obtain the numerical results. Then, studies are carried out to investigate the thermodynamic behavior of piles of different cross-sections (round pile of equal cross-section, cylindrical pile with a straight cone and cylindrical pile with a curved cone) in permafrost soils under freeze-thaw conditions.

In [7–9] physical processes occurring in the ground during freezing and thawing are modeled. A thermo-elastic-plastic model of unsaturated frozen ground based on a modified Cam-Clay model is proposed. This model effectively describes two main problems in the mechanical behavior of frozen soils: soil hardening caused by temperature decrease and volumetric compression during thawing. The proposed model is then implemented as part of a thermo-hydro-mechanical coupled model, the reliability of which is verified by comparison with experimental results.

The authors in [10] consider ice-water phase transitions, unsaturated conditions, and capillary action, and consider the effect of the developing pore volume on liquid-solid matrix interactions.

Thawing of ice-saturated rocks due to various anthropogenic impacts leads to subsidence of the earth surface and development of cryogenically dangerous geological processes called thermokarst. In [11] the results of numerical simulation of unsteady heat transfer processes in permafrost from heat sources generated by various technical systems (extraction wells), and from cold sources used for soil thermal stabilization (seasonal cooling units) are presented. The modeling takes into account average monthly fluctuations in temperature and solar radiation power typical of the geographical location under consideration, thermal parameters, and lithology of the ground as well as technical features of the engineering structures under consideration.

We can be concluded that at present there is no universal approach to the application of the system of continuum mathematical models, and non-redundant geometric models for solving the indicated problems. Most software products are focused on solving a limited set of model problems, and only some of them provide an opportunity for the user to change the mathematical models of the physical process, for example, ANSYS and COMSOL. Each change in the basic mathematical model leads to a change in the characteristics of the direct problem operator, which affects the properties of the discrete analog of the problem to be solved, regardless of the discretization method. Therefore, software developers are forced to use stable basic algorithms that are not specialized for a particular type of problem. There is a problem of choosing between multi-disciplinarity and orientation to a certain class of tasks in the development of scientific and technical software.

Numerical simulation of thermo-elastic deformation processes in frozen rock under cyclic changes in the temperature regime was performed in this work. We obtained that the greatest influence on the structural integrity of frozen rocks is exerted by production wells during the periods of maximum seasonal thawing.

The developed algorithms are specially designed for use in parallel computing systems. The algorithms are based on continuum models of multiphysics processes of permafrost rock evolution within a single mathematical and computational platform. This allows us to significantly increase the efficiency of solving these problems while maintaining the physical relevance of the numerical results.

2 Problem Statement

Figure 1 shows a geometric model of a frozen rock layer containing a layer of ice. The lateral dimension of the layer is 1 m. The thickness of the layer is 8 m. A well with a length of 5 m and a radius of 10 cm is driven into the bed. At a depth of 1 m, the well penetrates through a 1 m thick layer of ice.

A constant temperature of 20 °C is maintained at the surface of the well all year round. At the day surface, the temperature varies throughout the year according to the graph shown in Fig. 2.

The initial ground temperature is assumed to be –3 °C, specific heat of phase transition is 334 kJ/kg. The model time is one year. The time step in the calculations was one day.

Figure 3 shows the dependence of thermal conductivity of different types of parent rock skeletons on temperature.

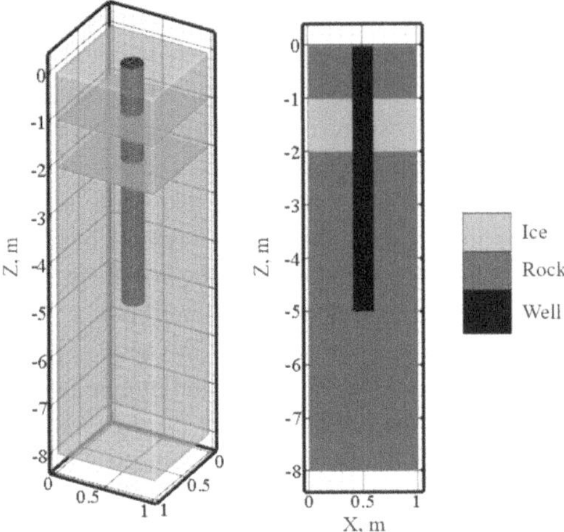

Fig. 1. Geometric model of frozen rock formation with a well

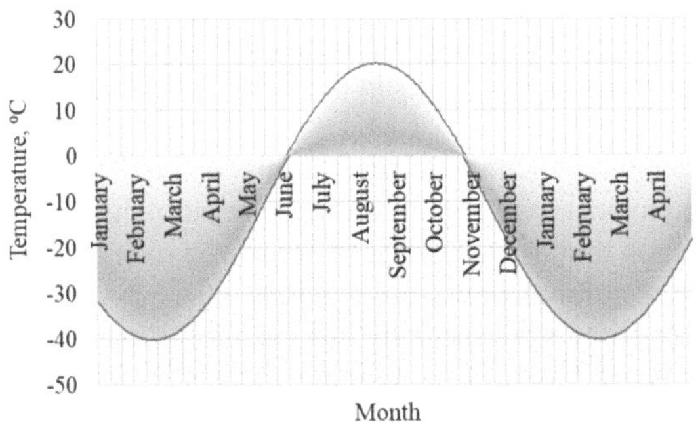

Fig. 2. Graph of seasonal temperature change day surface

Table1 contains dependences of thermal conductivity, specific heat capacity, and density of ice and water.

It is assumed that the specific heat capacity and skeleton density are not significantly affected by the ambient temperature. Table 2 presents the values of these characteristics that will be used in the calculations.

We consider a problem of numerical simulation of thawing and freezing in frozen rock taking into account natural and anthropogenic thermal effects, and the influence of the processes on the equilibrium properties of the rock.

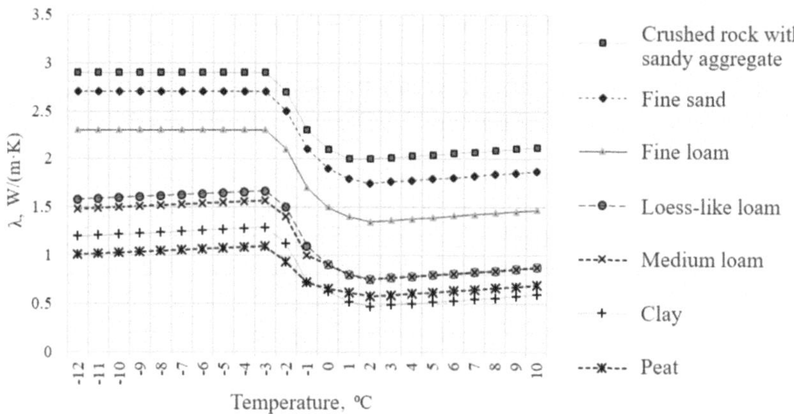

Fig. 3. Dependence of rock skeleton thermal conductivity on temperature

Table 1. Dependences of thermal conductivity, specific heat capacity, and density of ice and water

Temperature *T*, C	Thermal conductivity λ, W/(m·K)	Specific heat capacity c, J/(kg·K)	Density ρ, kg/m3
−40	2.6	1800	925
−30	2.5	1850	923
−20	2.4	1950	921
−10	2.3	2000	919
0	1.5	2100	918
10	0.6	4200	1000
20	0.65	4200	1000

Table 2. Density and specific heat capacity of reservoir layer components

	Crushed stone (5 m thickness)	Loam (thickness 15 m)	Clay (5 m thick)
Density ρ kg/m^3	2600	2740	2740
Specific heat capacity c, J/(kg·K)	835	900	900

3 Mathematical Model and Solution Method

3.1 Mathematical Model of Heat Transfer Process with Explicit Tracking of Phase Transformation Front and Its Discretization Method

Let Ω_i is a part of the computational domain filled with ice, Ω_w is a part of the computational domain filled with water, Ω_c is a part of the computational domain filled with the rock skeleton. We denote the boundary with constant temperature as S_1, insulated surface as S_2, and the boundary at which the heat exchange process with the environment takes place as S_3. Next, $\xi(t)$ is phase transition boundary.

The mathematical model of the heat conduction process with phase transition is described by the following system of differential equations:

$$\begin{aligned} \rho_i c_i \frac{\partial T_i}{\partial t} &= \nabla \cdot (\lambda_i \nabla T_i) \text{ in } \Omega_i, \\ \rho_c c_c \frac{\partial T_c}{\partial t} &= \nabla \cdot (\lambda_c \nabla T_c) \text{ in } \Omega_c, \\ \rho_w c_w \left(\frac{\partial T_w}{\partial t} + \mathbf{v} \cdot \nabla T_w \right) &= \nabla \cdot (\lambda_w \nabla T_w) + \mu \Psi \text{ in } \Omega_w, \end{aligned} \tag{1}$$

where ρ – density [kg/m^3], c – specific heat [J/kg · K], $\mathbf{v}$ – liquid phase velocity [m/s], λ – thermal conductivity [W/(m · K)], T – temperature [K], μ – dynamic viscosity [Pa·s], $\mu\Psi = 2\mu(\nabla^s \mathbf{v} : \nabla^s \mathbf{v})$– dissipation energy at viscous fluid flow [Pa/s].

Temperatures at the initial moment of time:

$$T_i|_{t=0} = T_i^0, \; T_c|_{t=0} = T_c^0, \; T_w|_{t=0} = T_w^0. \tag{2}$$

A constant temperature T_g (–3 ^{0}C in the problem) is maintained at the lower boundary S_1 of the parent rock:

$$T_c|_{S_1} = T_i|_{S_1} = T_g. \tag{3}$$

Boundary conditions for the system of Eqs. (1) on the thermally insulated side surface S_2 are determined as:

$$\lambda_i \nabla T_i \cdot \mathbf{n}|_{S_2} = 0, \; \lambda_c \nabla T_c \cdot \mathbf{n}|_{S_2} = 0, \; \lambda_w \nabla T_w \cdot \mathbf{n}|_{S_2} = 0. \tag{4}$$

There is a process of heat exchange with the environment on the surface S_2 with temperature T_{out}:

$$\begin{aligned} \lambda_i \nabla T_i \cdot \mathbf{n}|_{S_3} + \beta\left(T_i|_{S_3} - T_{\text{out}}\right) &= 0, \\ \lambda_c \nabla T_c \cdot \mathbf{n}|_{S_3} + \beta\left(T_c|_{S_3} - T_{\text{out}}\right) &= 0, \\ \lambda_w \nabla T_w \cdot \mathbf{n}|_{S_3} + \beta\left(T_w|_{S_3} - T_{\text{out}}\right) &= 0, \end{aligned} \tag{5}$$

where β – heat transfer coefficient [W/(K·m^2)].

Table 3. Heat transfer coefficient during the year

	I	II	III	IV	V	VI	VII	VIII	IX	X	XI	XII
β	0.65	0.56	0.6	0.8	13.6	20.3	20.8	19.8	13.8	3.8	1.67	0.85

The heat transfer coefficient β varies throughout the year at different natural temperature regimes. Table 3 presents this coefficient during the year.

On the phase transition boundary, ideal contact conditions with respect to temperature and inhomogeneous coupling conditions for heat flux (Stefan conditions) are satisfied:

$$[T]|_{\xi(t)} = 0, \tag{6}$$

$$[\lambda \nabla T \cdot \mathbf{n}]|_{\xi(t)} = -\rho H \frac{\partial \xi(t)}{\partial t}, \tag{7}$$

where H – specific enthalpy of phase transition [J/kg], ρ – density.

The specific enthalpy of phase transition for the solid phase is calculated by the formula:

$$H_i = \mathrm{c}_i T_i \tag{8}$$

and in a liquid phase:

$$H_w = L + \mathrm{c}_w T_w, \tag{9}$$

where c – specific heat capacity, T – temperature, L – specific heat of phase transition [J/kg].

To approximate the time derivative, we use the second-order Runge-Kutta scheme:

$$\begin{aligned} T_1^h &= T_k^h + \Delta t L_h\left(T_k^h\right), \\ T_2^h &= T_1^h + \Delta t L_h\left(T_1^h\right), \\ T_{k+1}^h &= \left(T_k^h + T_2^h\right)/2, \end{aligned} \tag{10}$$

where Δt – time step at the k-th iteration, $L_h\left(T^h\right)$ – spatial approximation.

On a non-empty set $\Omega \subset \mathbb{R}^3$ we consider a finite union of subsets $M_h(\Omega) = \bigcup_i K_i$ and the space $\Im_n(K_i)$ of polynomials of degree n. On the union $M_h(\Omega)$ we introduce a discrete function space to approximate the temperature:

$$Q^h = \left\{ q^h | q^h \in L_0^2(\Omega) : q^h \in \Im_n(K_i) \forall K_i \in M_h(\Omega),\ q^h \Big|_{\partial K_i} = 0 \right\}. \tag{11}$$

For the spatial discretization of the heat transfer problem based on the computational scheme of the discontinuous Galerkin method, we define the traces of functions on the external and internal boundaries $\Gamma = \bigcup_i \partial K_i$. On the set Γ we define the trace space $\mathrm{Tr}(\Gamma) = \prod_{\Omega_i \in M_h(\Omega)} L^2(\partial K_i)$. Let's denote multiple internal boundaries as $\Gamma_0 = \Gamma \backslash \partial\Omega$.

It is convenient to express the traces of ambiguous functions on interfragmentary boundaries through the averages $\{\cdot\}$ and jumps $[\cdot]$. For functions $\nabla T \in [\mathrm{Tr}(\Gamma)]^3$ and $T \in \mathrm{Tr}(\Gamma)$ averages $\{\cdot\}$ and jumps $[\cdot]$ on the outer boundaries $\partial\Omega$ are determined as:

$$\begin{gathered}[T]|_{\partial\Omega} = T\mathbf{n},\ \{T\}|_{\partial\Omega} = T, \\ [\nabla T]|_{\partial\Omega} = \nabla T \cdot \mathbf{n},\ \{\nabla T\}|_{\partial\Omega} = \nabla T,\end{gathered} \tag{12}$$

on the internal boundaries $\Gamma_0 = \partial K_i \cap \partial K_j$ we use the formulas:

$$\begin{gathered}[T]|_{\Gamma_0} = T_i\mathbf{n}_i + T_j\mathbf{n}_j,\ \{T\}|_{\Gamma_0} = \left(T_i + T_j\right)/2, \\ [\nabla T]|_{\Gamma_0} = \nabla T_i \cdot \mathbf{n}_i + \nabla T_j \cdot \mathbf{n}_j,\ \{\nabla T\}|_{\Gamma_0} = \left(\nabla T_i + \nabla T_j\right)/2.\end{gathered} \tag{13}$$

The variational formulation of the discontinuous Galerkin method for the problem of heat conduction in a liquid phase is as follows: to find $T_k^h \in Q^h$, that $\forall q^h \in Q^h$:

$$\begin{aligned}L_h\left(T_k^h\right) = \rho c \int_\Omega \mathbf{v} \cdot \nabla T_k^h q^h d\Omega + \int_\Omega \lambda \nabla T_k^h \cdot \nabla q^h d\Omega + \\ + \int_{\Gamma_0} \lambda\left(\left[T_k^h\right] \cdot \left\{\nabla q^h\right\} + \left\{\nabla T_k^h\right\} \cdot [q^h] + \tau[T_k^h] \cdot [q^h]\right) dS + \int_{S_3} \lambda\beta T_k^h\left(\mathbf{n} \cdot \nabla q^h\right) dS + \\ + \int_{\partial\Omega} \lambda\left(\left(\nabla T_k^h \cdot \mathbf{n}\right) v^h + \tau T_k^h q^h\right) dS - \int_\Omega \left(\mu\Psi^h + \frac{\rho c}{\Delta t} T_{k-1}^h\right) q^h d\Omega - \\ - \int_{S_1} \lambda T_g^h\left(\mathbf{n} \cdot \nabla q^h + \tau q^h\right) dS - \int_{S_3} \beta\lambda T_{\mathrm{out}}^h\left(\mathbf{n} \cdot \nabla q^h + \tau q^h\right) dS = 0,\end{aligned} \tag{14}$$

where $\tau = \rho\|\mathbf{v}\|/\lambda$.

To obtain a variational formulation of the discontinuous Galerkin method for the problem of heat conduction in solid phase and rock, we put $\mathbf{v} = \mathbf{0}$ and $\tau = \rho/\lambda$ in the formula (14).

The Stefan's condition (7) can be taken into account as:

$$\int_{\xi(t)} \lambda\left\{\nabla T^h\right\} \cdot [\psi^h] dS = \int_{\xi(t)} \frac{\lambda^s \nabla T^s + \lambda^l \nabla T^l}{2} \cdot \psi^h \mathbf{n} dS = -\int_{\xi(t)} \frac{1}{2} \rho H \mathrm{V} \psi^h dS, \tag{15}$$

where V is the velocity of phase transition front motion.

The first order hierarchical basis of the $\mathrm{H}_1(\Omega)$ space is used to approximate the temperature field. A more detailed description of the solution method and its verification are given in [12].

3.2 Mathematical Model of Heat Transfer Process with Phase Transformations in Enthalpy Approach and Its Discretization Method

We use the already introduced notations from Sect. 3.1. The mathematical model of the heat conduction process with phase transition in frozen ground is described by the following system of differential equations:

$$\left(\rho_i c_i + \frac{\rho_w L}{T^l - T^s}\right)\frac{\partial T_i}{\partial t} = \nabla \cdot (\lambda_i \nabla T_i) \text{ in } \Omega_i,$$
$$\rho_w c_w\left(\frac{\partial T_w}{\partial t} + \mathbf{v} \cdot \nabla T_w\right) = \nabla \cdot (\lambda_w \nabla T_w) + \mu\Psi \text{ in } \Omega_w, \tag{16}$$

where ρ – density [kg/m^3], c – specific heat [J/kg · K], $\mathbf{v}$ – liquid phase velocity [m/s], λ – thermal conductivity [W/(m · K)], T – temperature [K], μ – dynamic viscosity [Pa·s], $\mu\Psi = 2\mu(\nabla^s\mathbf{v} : \nabla^s\mathbf{v})$– dissipation energy at viscous fluid flow [Pa/s], T^l – liquidus temperature [K], T^s – solidus temperature [K], λ – thermal conductivity [W/(m · K)].

Temperatures at the initial moment of time:

$$T_i|_{t=0} = T_i^0, \; T_w|_{t=0} = T_w^0 \tag{17}$$

A constant temperature T_g (–3 ^{0}C in the problem) is maintained at the lower boundary S_1 of the parent rock:

$$T_c|_{S_1} = T_i|_{S_1} = T_g. \tag{18}$$

Boundary conditions on the thermally insulated side surface S_2 are as follows:

$$\lambda_i \nabla T_i \cdot \mathbf{n}|_{S_2} = 0, \; \lambda_w \nabla T_w \cdot \mathbf{n}|_{S_2} = 0. \tag{19}$$

There is a process of heat exchange with the environment on the surface S_2 with temperature T_{out}:

$$\lambda_i \nabla T_i \cdot \mathbf{n}|_{S_3} + \beta\left(T_i|_{S_3} - T_{\text{out}}\right) = 0,$$
$$\lambda_w \nabla T_w \cdot \mathbf{n}|_{S_3} + \beta\left(T_w|_{S_3} - T_{\text{out}}\right) = 0, \tag{20}$$

where β – heat transfer coefficient [W/(K·m^2)].

The discretization method presented in Sect. 3.1 can be applied to this problem formulation. To obtain a variational formulation of the discontinuous Galerkin method for the problem of heat conduction in frozen ground, we put $\mathbf{v} = \mathbf{0}$, $\rho c = \rho_i c_i + \rho_w L/\left(T^l - T^s\right)$, and $\tau = \rho/\lambda$ in the formula (14).

3.3 Mathematical Model of Mechanical Deformation Due to Change in Temperature and Its Discretization Method

We formulate a mathematical model describing the thermoelastic deformation of the solid part of the rock (matrix Ω_0). The defining equations in this case are the heat conduction and balance equations (Hooke's law):

$$c_V \rho \frac{\partial T}{\partial t} = \nabla \cdot \lambda \nabla T \text{ in } \Omega_0, \tag{21}$$

$$\rho\frac{\partial^2\mathbf{u}}{\partial t^2} = \nabla\cdot\mathbf{D}(t) : \left(\nabla^S\mathbf{u} - \alpha\left[T - T_{k-1}\right]\right) + \mathbf{F}\,\text{in}\,\Omega_0, \tag{22}$$

where t – time [s], c_V – heat capacity at constant volume [J/kg·K], T – temperature [^{0}C], λ – thermal conductivity [W/m · K], ρ – density [kg/m^3], $\mathbf{u} = \left(u_x, u_y, u_z\right)^T$ – displacement vector [m], $\boldsymbol{\sigma}$ – stress tensor [Pa], $\mathbf{D}(t)$ – elastic tensor (Hooke's tensor) [Pa], $\nabla^S(\cdot)$ – symmetrical part of the gradient, $\boldsymbol{\alpha}$ – thermal expansion tensor [K^{-1}], $\mathbf{F}$ – gravity [Pa/m].

The following boundary conditions are introduced to couple the temperature fields on the pore surfaces $\partial\Omega_{in}$:

$$\lambda\nabla T(\partial\Omega_{in})\cdot\mathbf{n} + \beta(T - T(\partial\Omega_{in})) = 0, \tag{23}$$

where β – heat transfer coefficient [W/m^2·K] $\mathbf{n}$– external unit normal to the considered boundary.

On the day part of the ground surface fragment under consideration, the temperature is determined by some function of time $T_{up}(t)$. The lower boundary of the region is considered to be remote with constant temperature T_0 and no displacements. On the lateral boundaries, the conditions of thermal insulation and free displacements are specified:

$$T(z = z_{\max}, t) = T_{up}(t) \tag{24}$$

$$T(z = z_{\min}, t) = T_0, \tag{25}$$

$$\frac{\partial T(z \neq \{z_{\min}, z_{\max}\}, t)}{\partial\mathbf{n}} = 0, \tag{26}$$

$$\mathbf{u}(z = z_{\min}, t) = \mathbf{0}, \tag{27}$$

$$\boldsymbol{\sigma}(z \neq z_{\min}, t)\cdot\mathbf{n} = \mathbf{0}. \tag{28}$$

At the initial moment of time, the temperature field is homogeneous, and the whole thermomechanical system is in equilibrium state:

$$T(z, t = 0) = T_0, \tag{29}$$

$$\mathbf{u}(z, t = 0) = \mathbf{0}, \tag{30}$$

$$\frac{\partial\mathbf{u}(z, t = 0)}{\partial t} = \mathbf{0}. \tag{31}$$

For the numerical solution of the initial boundary value problems, a hierarchical finite element discretization and the corresponding discrete variational formulation of the heterogeneous finite element method are constructed. The solution thus obtained is transferred at each time step through conjugation conditions at the interfaces to solve the problems formulated in pore space Ω_{in}.

Thus, the discrete variational formulation for the thermoelastic deformation problem in the ideology of the heterogeneous multiscale finite element method is as follows: find $(\mathbf{u}, T)^H \in \mathbf{V}^H(\Pi^H(\Omega)) \cup V^H(\Pi^H(\Omega)) + (\mathbf{u}, T)(\partial\Omega)$, that $\forall v^H \in V^H(\Pi^H(\Omega))$ and $\forall \mathbf{v}^H \in \mathbf{V}^H(\Pi^H(\Omega))$:

$$\begin{gathered}\int\limits_{\Omega} \nabla \mathbf{u}^H : \mathbf{D}^H : \nabla \mathbf{v}^H d\Omega = \int\limits_{\Omega} \rho \frac{\partial^2 \mathbf{u}^H}{\partial t^2} \mathbf{v}^H d\Omega + \int\limits_{\Omega} \mathbf{F} \mathbf{v}^H d\Omega + \\ + \int\limits_{\Omega} \mathbf{v}^H \cdot \nabla \cdot \left(\mathbf{D}^H : \alpha\left[T^H - T_{k-1}\right]\right) \mathbf{v}^H,\end{gathered} \tag{32}$$

$$\begin{gathered}\int\limits_{\Omega} \lambda \nabla T^H \nabla v^H d\Omega + \int\limits_{\partial\Omega_{in}} \left(\lambda \nabla T^H \cdot \mathbf{n} + \beta T^H\right) v^H d(\partial\Omega) = \\ = \int\limits_{\Omega} c_V \rho \frac{\partial T^H}{\partial t} v^H d\Omega + \int\limits_{\partial\Omega_{in}} \beta T_{out} v^H d(\partial\Omega),\end{gathered} \tag{33}$$

where $\mathbf{D}^H$ – homogenized media coefficient, $\mathbf{V}^H$ and V^H – discrete Hilbert subspaces of a Hilbert subspace defined as a linear envelope over the set of finite nonpolynomial form functions constructed on a macro-element polyhedral mesh $\Pi^H(\Omega)$.

4 Results

4.1 Influence of Seasonal Change of Temperature Background During the Annual Period and Anthropogenic Effects Associated with Well Operations on the Process of Permafrost Degradation

Now, we present computation results of solving the problem described in Sect. 2. Figure 1 shows a geometric model of a frozen rock layer containing a layer of ice. The lateral dimension of the layer is 1 m. The thickness of the layer is 8 m. A well with a length of 5 m and a radius of 10 cm is driven into the bed. At a depth of one meter, the well penetrates through a 1 m thick layer of ice.

A constant temperature of 20 ^{0}C is maintained on the surface of the well all year round. On the day surface, the temperature varies throughout the year according to the graph shown in Fig. 2. The initial ground temperature is assumed to be –3 ^{0}C, specific heat of phase transition is 334 kJ/kg. The model time is one year. The time step in the calculations was one day.

Table 1 contains dependences of thermal conductivity, specific heat capacity, and density of ice and water. The porosity of the parent rock to a depth of 5 m is assumed to be $\varphi = 10\%$, which corresponds to the coarse-grained fraction. The horizon from 5 m to 8 m has porosity $\varphi = 30\%$, which corresponds to loam. To estimate the effective physical characteristics of the parent rock, we apply relations:

$$\begin{aligned}\lambda_{\text{melt rock}} &= (1-\varphi)\lambda_{\text{rock skeleton}}(T) + \varphi\lambda_{\text{water}}(T),\\ \lambda_{\text{frozen rock}} &= (1-\varphi)\lambda_{\text{rock skeleton}}(T) + \varphi\lambda_{\text{ice}}(T),\end{aligned}$$

$$(\rho c)_{\text{melt rock}} = (1-\varphi)(\rho c)_{\text{rock skeleton}}(T) + \varphi(\rho c)_{\text{water}}(T),$$
$$(\rho c)_{\text{frozen rock}} = (1-\varphi)(\rho c)_{\text{rock skeleton}}(T) + \varphi(\rho c)_{\text{ice}}(T). \quad (34)$$

It is assumed that the specific heat capacity and density of the skeleton are not significantly affected by the ambient temperature. Table 2 shows the values of these characteristics that will be used in the calculations.

The problem of numerical simulation of the process of thawing and freezing of frozen rock taking into account anthropogenic thermal effects and the influence of the thawing process on the equilibrium properties of the rock is posed.

The calculations were performed using AMD Ryzen 9 3950X 16-Core Processor 3.49 GHz, 64 GB GPU, Windows 10 (×64). The computational mesh contains 392476 tetrahedrons.

Figure 4 shows the state of the frozen soil model and the change in temperature over a one-year cycle from January to January of the following year, assuming a constant temperature at the top surface of the well.

During winter and early summer, the presence of a permanent heat source in the form of a well leads to thawing of the upper frozen soil layers up to 4 cm from the well surface. By mid-July, the daytime temperature reaches 20 ^{0}C, and there is no effect of the well on the thawing of the upper frozen rock layers. Now, the phase transition process in deeper horizons of the formation is initiated only due to the temperature increase at the day surface. During this period, the phase transition front reaches the ice layer. It leads to its melting. By mid-September, the melting boundary reaches a depth of 3 m.

The melting of deep ice under seasonal temperature changes actively begins in July and ends in September. Figure 4 shows the melting ice layer in black color.

In October, negative temperatures are established at the day surface and two fronts of freezing appear. The first front spreads downward from the day surface. The second front spreads upward from deeper frozen ground layers. As can be seen in Fig. 6, positive temperatures persist only at the depth of the melted ice layer. By mid-November, this area freezes again. In winter, the presence of a well with a constant temperature also leads to the formation of a phase transition region 4 cm thick from its surface.

Since the discontinuous Galerkin method is used for space approximation, the resulting system of algebraic equations will have a block structure. In addition, the time approximation is performed using the Runge-Kutta computational scheme. In this regard, the resulting system of algebraic equations will have a block diagonal structure. Therefore, to solve the problem described, we can effectively apply parallelization techniques.

In Table 4, we demonstrate the time of solving the problem with increasing number of physical CPU cores. The linear decrease of the problem-solving time indicates the natural parallel structure of the developed algorithm.

Table 4. Time of solving the problem with increasing number of physical CPU cores

CPU cores	1	2	4	8	16
Time, s	216.0	102.9	55.4	27.7	13.5

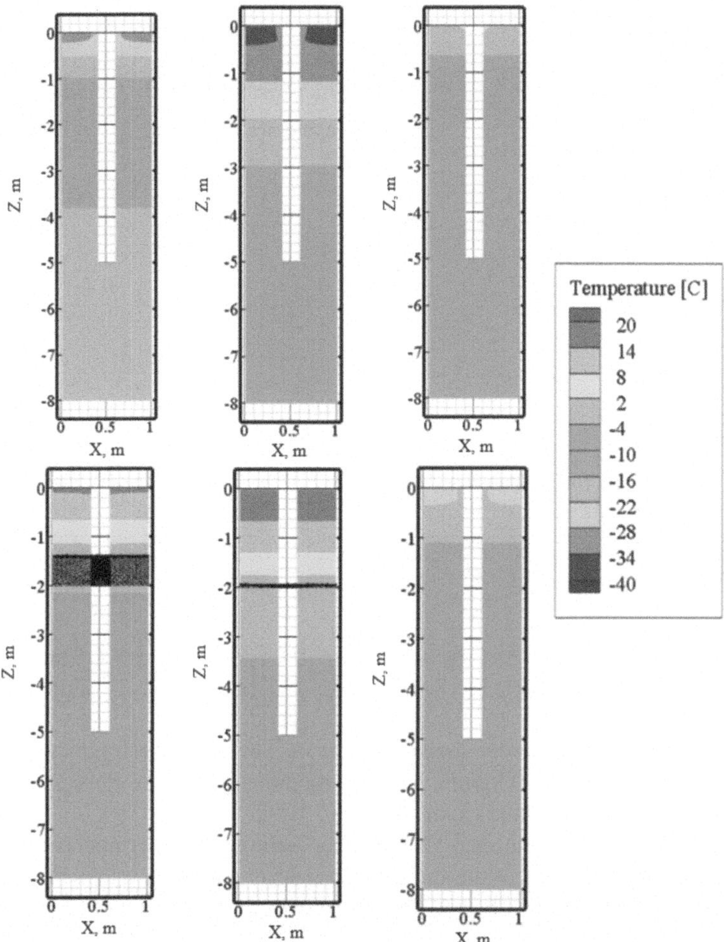

Fig. 4. Frozen ground conditions and temperature front in January, in March, in June, in July (beginning of ice layer melting), in September (completion of ice layer melting), in December

4.2 Mechanical Deformation of Permafrost Rocks Taking into Account Cyclic Temperature Changes

At the initial moment of time the temperature in the whole area is assumed to be –3 °C, then the seasonal temperature change at the day surface of the reservoir is simulated using the function According to the change in the temperature field there is a change in the physical properties of the components of the region as shown in Table 5.

Table 5. Some physical characteristics at different temperature

	Dusty permafrost soil		Fluid in solid phase
	Temperature, ^{0}C	Value	
Young's modulus, GPa	−7	1.96	3
	0	0.7	
	3	0.05	
Poisson's ratio	−4.0	0.13	0.4
	−1.5	0.14	
	−0.8	0.18	
	−0.3	0.35	
	3	0.45	
Coefficient of linear thermal expansion, $^0C^{-1}$	−9	88E+5	0.158E−3
	−7	70E+5	
	−5	25E+5	
	−3	22E+5	
	3	18E+5	
Density, kg/m^3	–	2650	920

We use the geometric model showed in Fig. 4. The reservoir fragment contains a cased well with density 2500 kg/m^3, Young's modulus 32.5 GPa, Poisson's ratio 0.2, and coefficient of linear thermal expansion 1E+5 °C^{-1}.

Figure 5 shows computational results after solving the thermoelastic deformation problem. When the ice in the interbed space has completely melted, the largest deformation (about 10 cm) is observed after 274 days from the start of observations.

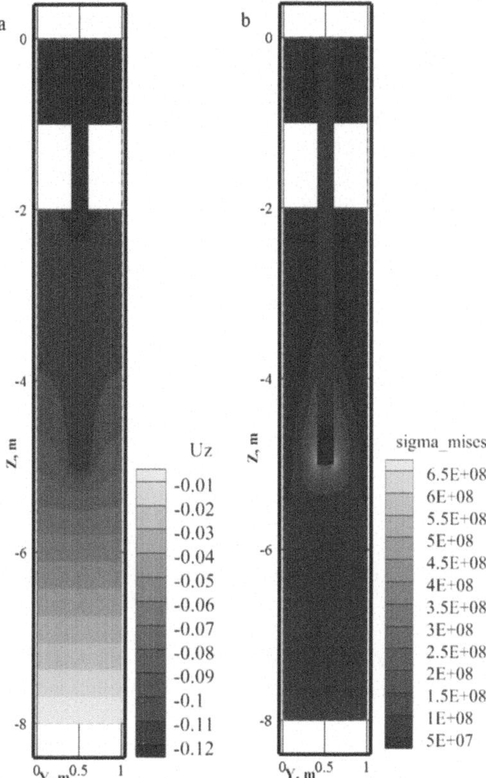

Fig. 5. Distribution of Z-component of displacement vector (a) and von Mises stress (b) in section X = 0.5 m after 274 days from the start of observations

The computational experiment results have been obtained, taking into account the force of gravity. We can conclude, thawing occurs at a depth of no more than 5 m at the seasonal temperature fluctuations considered, while the deformation of the formation is no more than 2 cm.

5 Conclusions

The problem of permafrost layer degradation under external natural and anthropogenic temperature effects was considered. The problem of thawing of downhole space was solved taking into account seasonal temperature changes during the year, and taking into account the influence of the cased well temperature. A multiscale mathematical model of the heat transfer process based on the Stefan problem with explicit tracking of the phase transition front and in enthalpy formulation was proposed. The process of thermoelastic deformation of the rock was modeled taking into account changes in the physical properties of the thawed soil, and the influence of gravity. Computational schemes of non-conforming finite element methods were used to discretize the mathematical models. These schemes in cooperation with the Runge-Kutta schemes for time discretization allowed to implement naturally parallel algorithms.

The obtained results of computational experiments allow us to draw the following conclusions. In the periods of negative temperatures, the process of thawing is determined only by the temperature value of the wellbore. The maximum lateral depth of thawing can reach 5 cm. In periods of positive temperatures, the thawing reaches a depth of 5 m. During a calendar year, the processes of ground thawing and freezing lead to a maximum deformation up to 2 cm, which can be a fatal factor for the stability of technical structures in permafrost areas.

References

This work was carried out with the financial support of the RSF (project No. 25-71-10045).

References

1. Golik, V.V., Moiseev, B.V., Gulkova, S.G., Zemenkov, Y.D.: Mathematic simulation of the effect of a buried oil pipeline on perma-frost soils. IOP Conf. Series: Mater. Sci. Eng. **445**, 012004 (2018)
2. Golik, V.V., Moiseev, B.V., Zemenkov, Y.D., Kabes, E.N.: Assessment of geo-cryological conditions in the design and operation of pipelines in the Arctic zone of the Russian Federation. IOP Conf. Series: Mater. Sci. Eng. **445**, 012005 (2018)
3. Golik, V.V., Zemenkov, Yu.D., Shagbanov, I.F.: Complex thermophysical modeling of processes in the foundation soil of oil pipelines in the Arctic and offshore conditions. IOP Conf. Series: Mater. Sci. Eng. **952**, 012014 (2020)
4. Zhou, J., Zhang, L., Liang, Z., Zheng, T., Zheng, J.: Thermo-mechanical stability analysis of thermosyphons for pipeline soil and their optimal layout in permafrost regions. Cold Regions Sci. Technol. **218**, 104078 (2024)
5. Chen, P., et al.: Development of safe operation technology of crude oil pipe-line in permafrost regions. J. Pipeline Sci. Eng. **4(2)**, 100152 (2023)
6. Akimova, E.N., Filimonov, M.Yu., Misilov, V.E., Vaganova, N.A.: Simulation of thermal processes in permafrost: parallel implementation on multicore CPU. CEUR Workshop Proc. **2274**, 1–9 (2018)
7. Zhou, Y., et al.: Calculation method and model tests of pile frost jacking for railway overhead contact systems in permafrost regions. Cold Regions Sci. Technol. **206**, 103746 (2023)
8. Liu, Q., Cai, G., Zhou, C., Yang, R., Li, J.: Thermo-hydro-mechanical coupled model of unsaturated frozen soil considering frost heave and thaw settlement. Cold Regions Sci. Technol. **217**, 104026 (2024)
9. Ma, M., Chen, H., Ma, Y.: Numerical evaluation of thermal stability of a W-shaped crushed-rock embankment with shady and sunny slopes in warm permafrost regions. Int. Commun. Heat Mass Transfer **144**, 106754 (2023)
10. Jonas, B., Shubhangi, G., Barbara, W., Gabriele, C.: Alpine permafrost modeling: on the influence of topography driven lateral (2021). https://arxiv.org/abs/2110.07217
11. Filimonov, M.Y., Vaganov, N.A.: Simulation of thermal stabilization of soil around various technical systems operating in permafrost. Appl. Math. Sci. **7(144)**, 7151–7160 (2013)
12. Markov, S.I.: A discontinuous Galerkin method for mathematical modeling of icemelting at the interaction with the environment. IOP Conf. Ser.: Earth Environ. Sci. **193**, 012043 (2018)

Development of a Simulation and Laboratory Test Bench for Studying the Energy Characteristics of a Micro Hydroelectric Power Plant

Ishembek Kadyrov[1(✉)], Taalay Mederov[2], Bermet Zhanybekova[1], and Maksat Narymbetov[1]

[1] Skryabin Kyrgyz National Agrarian University, Mederov Street 68, 720005 Bishkek, Kyrgyz Republic
bgtu_kg@mail.ru
[2] Razzakov Kyrgyz State Technical University, Ch. Aitmatov Avenue 66, 720044 Bishkek, Kyrgyz Republic

Abstract. The paper addresses the need for alternative energy sources to supply decentralized regions of Kyrgyzstan—particularly farming communities and geological survey teams—with reliable electric and thermal power. Given that over 80% of the country's territory is mountainous, utilizing the energy of mountain rivers through the development of small or micro hydropower plants (HPPs) is a national priority. A central challenge in designing such systems is the selection of an appropriate hydroturbine, as this choice directly impacts the efficient use of the fluctuating water flow in mountainous terrain. Seasonal and climatic variations cause changes in water head and flow rate, making it difficult to maintain the turbine's rated power output throughout the year. To explore effective ways of adapting turbine performance to these variable conditions, the study proposes a simulation-based laboratory setup. The experimental stand is built around a pair of AC machines: an asynchronous (induction) motor, which simulates the mechanical power generation from a hydroturbine, and a synchronous generator, which models the electrical load using a combination of active and reactive components. The turbine's mechanical characteristics are derived through CFD modeling using FlowVision software, which solves the Navier–Stokes and continuity equations to determine turbine head as a function of power output. These data are then transformed into universal turbine performance curves used to replicate turbine behavior in the lab. Variations in river head or flow rate are simulated via a frequency converter and nonlinear control block, enabling the induction motor to imitate changes in turbine speed and output power. The system's built-in PI controllers ensure stable operation across different operating conditions. This approach provides a practical framework for evaluating turbine designs under real-world variability in mountain water resources, supporting more efficient deployment of micro-HPPs in remote and energy-scarce regions.

Keywords: Hydroturbine Simulation · Frequency Converter · Induction Motor · FlowVision · Tiristor Converter · Micro Hydropower Plant · Renewable Energy

V. Jordan et al. (Eds.): HPCST 2024, CCIS 2919, pp. 85–99, 2026.
https://doi.org/10.1007/978-3-032-20325-0_7

1 Introduction

The main sources of electrical energy in Kyrgyzstan are large hydropower plants (HPPs), which generate approximately 90% of the country's electricity. In contrast, thermal power plants powered by hydrocarbon fuels produce less than 10%. These centralized facilities serve about 20% of the national territory, primarily densely populated regions. The remaining territory consists mostly of mountainous areas, where small, dispersed settlements fall under the category of decentralized regions. Extending power lines or transporting hydrocarbon fuels to these remote areas is economically inefficient. As a result, the only viable option for supplying electrical and thermal energy is through alternative energy sources – namely, wind, solar, and the energy of small mountain rivers. [1, 2].

The estimated hydropower potential of Kyrgyzstan's small rivers and streams is around 1.6 million kW, with an annual energy production capacity of 5 to 8 billion kWh. In recent years, national energy policy has increasingly focused on expanding the use of alternative energy sources. [1] The goal is to meet the needs of both centralized and decentralized areas while ensuring long-term energy security.

The emphasis on renewable energy – particularly small mountain rivers and solar energy – is driven by the high cost of imported fuel and the logistical challenges of fuel delivery to remote regions. In these locations, alternative energy solutions are implemented through the construction of small HPPs that supply electricity, hot water, and heating. [2] This is especially important for rural communities, farms, geological field teams, and other small-scale operations in hard-to-reach mountainous areas.

Kyrgyzstan's abundance of mountain rivers offers a unique opportunity to harness renewable resources by developing both small and micro HPPs. The choice between them depends on two main factors: the required power output and the local terrain. For higher energy demand, such as in farming settlements, a small diversion-type HPP should be constructed, taking advantage of the natural landscape and including all necessary infrastructure. For lower demand applications, such as powering geological teams or beekeepers, a micro HPP utilizing the natural head of a mountain stream is sufficient. [3].

In either case, selecting the appropriate turbine is a critical decision. The turbine is the key component that converts the energy of mountain streams into mechanical energy, and its design significantly affects efficiency and cost.

Various types of micro HPPs are used worldwide to power autonomous, decentralized consumers. These systems differ in turbine design, and while many offer technical advantages, they often suffer from drawbacks such as excessive size, high material consumption, and cost. Therefore, a key research challenge is the development of new, highly efficient, reliable, and economically feasible micro HPP designs. [3].

This article aims to present the development of a simulation-based laboratory test stand that enables designers to study the performance of a selected turbine configuration early in the design process. This approach helps identify turbine designs that are well-suited to the micro HPP structure while avoiding common technical limitations.

2 Principles of Designing the Simulation-Based Laboratory Test Stand

The designed laboratory test stand is not intended for studying the structural features of hydroturbines, but rather for investigating the effectiveness of their application once a mathematical description or another form of representation of the turbine's mechanical characteristics is determined.

In practice, the process of selecting key components for a micro HPP begins with the dilemma of choosing the appropriate turbine design. This decision is crucial as the turbine is the main component that converts the energy of mountain rivers into mechanical energy. Moreover, the performance of the micro HPP is directly influenced by the design of the hydroturbine, considering the variability in water flow through the turbine blades, which depends on seasonal changes and local environmental conditions. Therefore, the primary task is to study the performance of the micro HPP and select the optimal turbine design. [4].

To achieve this goal, the designer initially selects the turbine design using manufacturer recommendations before starting the micro HPP design process. In cases where the manufacturer's technical documentation lacks a mechanical characteristic for the selected turbine design, laboratory research is required to study its energy capabilities. This involves creating a simulation model and formulating recommendations for its application.

The simplest way to create a laboratory simulation test stand is by using a pair of interconnected alternating current machines: a synchronous generator and an induction motor. The analysis shows that the key characteristics of the power equipment in a micro HPP can be determined if the main parameters of the mountain river, such as head (H, in meters) and flow rate (Q, in m^3/s), are applied to the induction motor. Since these parameters directly affect the turbine's angular speed (ω, in rad/s), and consequently the power generated by the hydroturbine, obtaining data on the turbine's Q-H characteristics at an early stage of micro HPP design becomes an essential task. The simulation test stand allows the initial head H to be set according to the local terrain features and maintained at a constant value (H = constant) during the tests.

From this, it follows that in the simulation test stand, the induction motor (IM) should closely replicate the operation of the hydroturbine. If this goal is achieved, the induction motor will exhibit the mechanical characteristics of the hydroturbine, allowing for the determination of the turbine's output power at any speed, which depends on variations in the water flow Q.

Another important parameter for the object under study is the simulation of the input parameter of the hydroturbine, i.e., the change in water flow Q, which affects the turbine's rotational speed. According to electric drive theory, the speed of an induction motor (IM) can be controlled over a wide range using a frequency converter (FC). [5] Therefore, the frequency converter (FC) as part of the automation system should simulate the variability of water flow, which depends on both seasonal changes and local climatic conditions. If the FC is set to operate in current source mode, this goal can be achieved with simple means.

Thus, the electric drive system consisting of a frequency converter and induction motor (FC-IM) can simulate the operation of the designed small and micro HPPs, provided that the mechanical characteristics of existing hydroturbine models can be determined through their parameters, taking into account the design features and inputting these parameters into the induction motor.

2.1 Functionality of the Control Panel

The functional diagram of the simulation stand, corresponding to the above-described processes of converting the energy of mountain rivers into electricity, is shown in Fig. 1. In this diagram, the main element of the micro HPP – the hydroturbine – is represented by the mechanical characteristics of the electrical drive system in the FC-IM setup, operating in a closed speed control system. The characteristic labeled as $\omega_{ht} = f(M_{ht})$, constructed from points labeled 1, 2,..., 5, is the desired characteristic. If the characteristic $\omega_{ht} = f(M_{ht})$ can be interpreted, in other words, the *Q-H* characteristic of the chosen hydroturbine can be determined, then to set the operation of the IM at points 1, 2,..., 5, it is sufficient to introduce a nonlinear block (NB) into the electrical drive control system. The NB ensures the setting of the current to the frequency converter so that the motor torque corresponds to the currents supplied to the IM windings in accordance with the specified *Q-H* characteristic of the hydroturbine. Thus, the nonlinear block allows the formation of the mechanical characteristic $\omega_{ht} = f(M_c)$, inherent to the investigated turbine design, whose power depends on the water flow rate *Q*.

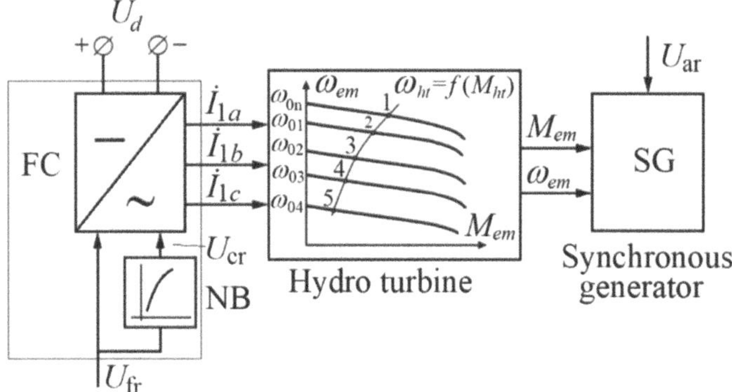

Fig. 1. Functional diagram of the simulation stand designed for studying energy processes in micro HPPs

The synchronous generator (SG) with a rated power of $P_{SG} = 3.7$ kW, as a physical model, is preserved in its natural form as part of the micro HPP simulation elements. The input parameters for the SG are the power from the hydroturbine supplied to the shaft of the SG, and the output parameter is the electrical energy produced by the generator. The adjustable parameter for the generator is the amplitude of the stator output voltage, controlled by the signal U_{ar}.

In the functional diagram shown in Fig. 1, the frequency converter forms the mechanical characteristics of the electrical drive with the specified synchronous rotational speeds $\omega_{0nom}, \omega_{01}, \ldots, \omega_{04}$. The NB determines the steady-state operating modes of the electrical drive with loads at the points labeled 1, 2,..., 5.

Thus, with high accuracy, it can be stated that the points labeled 1, 2,..., 5 are the simulation points for the power generated by the hydroturbine, which has the characteristic $\omega_{ht} = f\ (M_{ht})$.

To achieve accuracy in the operation of the electrical drive at the specified points, proportional-integral current regulators are used in the FC-IM control system. With the proper adjustment of these regulators, it is easy to ensure steady operating modes for the electrical drive if the frequency converter is set to the current source mode. The only remaining task is to determine the law of change for the curve enclosed in the NB, which in graphical or tabular form, at the appropriate scale, is entered as the experimentally calculated graphs of the mechanical characteristics $\omega_{ht} = f\ (M_{ht})$ [6].

Thus, the main task in achieving the stated goal of this article is to develop methods for calculating the mechanical characteristics of the selected hydroturbine design.

2.2 Data Preparation and Preliminary Calculations of the Axial Hydroturbine

As mentioned in the introduction, for enterprises located in decentralized areas, two types of micro-HPPs can be constructed, depending on their rated power. The first type is the diversion small HPP, which is equipped with a reactive hydraulic turbine. Small water reservoirs are built for these HPPs to accumulate potential and kinetic energy, and according to operational characteristics, reactive hydroturbines must be fully submerged in water.

The second type is the floating micro-HPP, which uses active hydroturbines to convert the kinetic energy of water into mechanical energy, placed in an open environment. It is worth noting that the hydraulic engine, the hydroturbine, gets its name because water is the primary energy source that transmits motion to the turbine's working wheel.

When designing a small HPP or micro-HPP, the main task is to select the necessary hydroturbine design based on the criterion of maximum power generation. This can be done using the map in Fig. 2, which highlights the zones for each type of hydroturbine, where they operate most efficiently, depending on the flow rate Q (m^3/s) and head H (m) of the water. As shown in Fig. 2, at low flow rates H and high heads Q, active bucket hydroturbines (Pelton) are most effective, whereas at lower heads and higher flow rates, adjustable blade, propeller-type hydroturbines (Kaplan) are preferred. Radial-axial hydroturbines (Francis) cover a larger range of zones, and so on.

The mechanical characteristic calculation will be carried out by first selecting the propeller-type hydroturbine design shown in Fig. 3, as it is simple to manufacture and operate, with minimal cost.

The mechanical characteristic for the hydroturbine shown in Fig. 3 will be calculated using the FlowVision software suite, preparing the initial data for computer modeling. [7] The main parameters for modeling the propeller wheel include: the diameter of the working wheel D, the hub diameter d, the number of working blades Z, and the blade profiles of the working wheel. These parameters are necessary for the calculation of the micro-HPP data, which include the generator bus power P_{GEN} (kW), the head of water

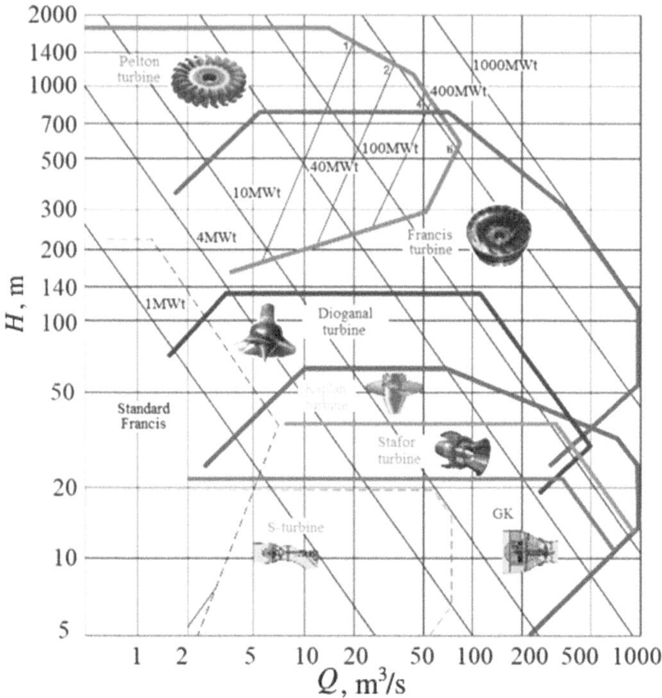

Fig. 2. Operating ranges of different types of hydroturbines

on the hydroturbine H (m), and the turbine shaft rotational speed n (rpm), prepared for input into the FlowVision software suite.

Fig. 3. Overall design of the propeller-type hydroturbine

The hydroturbine power N_{HT} is determined taking into account hydraulic k_h, mechanical k_m, and electrical losses k_e using the following formula:

$$N_{HT} = \frac{P_{GEN}}{k_h \cdot k_m \cdot k_e}. \tag{1}$$

Based on the calculated power value, we determine the water flow rate passing through the hydroturbine runner blades:

$$Q = \frac{N_{HT}}{9.81 \cdot H}. \tag{2}$$

Specific speed n_s of the designed hydroturbine:

$$n_s = \frac{n \cdot \sqrt{P_{HT}}}{H \cdot \sqrt[4]{H}}. \tag{3}$$

From which the outer diameter of the runner is:

$$D = \frac{6.7}{H \cdot \sqrt[3]{n_s + 100}} \cdot \frac{\sqrt{Q}}{\sqrt[4]{H}}. \tag{4}$$

The hub diameter is determined based on recommended hub-to-runner diameter ratios (HTR_0), obtained experimentally to optimize the energy performance of the runner under known head conditions.

$$d = HTR_0 \cdot D. \tag{5}$$

Number of blades:

$$Z = \frac{360}{\theta} \cdot \frac{L}{t}, \tag{6}$$

where θ is the blade angle, θ = 70 to 80° [8]; L / t is the average grid density.

3 CFD-Simulation

Computational Fluid Dynamics (CFD) is a branch of continuum mechanics that encompasses a set of physical, mathematical, and numerical methods used to calculate characteristics of flow processes. Computer simulations were conducted using the FlowVision software suite, which incorporates various turbulence models and adaptive mesh techniques. FlowVision allows for the simulation of complex fluid motion, including flows with significant swirling.

The primary equations for calculations are the Navier-Stokes equation along with the continuity equation. The Navier-Stokes equation describes the motion of viscous incompressible fluids [9].

$$\frac{d\overrightarrow{v}}{dt} = \overrightarrow{F} - \frac{1}{p}\mathrm{grad}\,p + \nu\nabla^2\overrightarrow{v}, \tag{7}$$

где $\overrightarrow{dv} \big/ dt$ – the total acceleration of a fluid particle; $\overrightarrow{F}$ – the acceleration due to body forces; $(1/p)$ *grad p* – the acceleration due to pressure forces; $\nu\nabla^2\overrightarrow{v}$ – the acceleration due to viscous forces.

In cylindrical coordinates, the aforementioned equation is written as

$$\begin{cases} \frac{\partial v_r}{\partial t} + v_r \frac{\partial v_r}{\partial r} + v_z \frac{\partial v_z}{\partial r} - \frac{v_\theta^2}{r} = 2\omega_0 v_\theta - \frac{1}{\rho}\frac{\partial \rho}{\partial r} + \nu\left(\frac{1}{r}\frac{\partial v_r}{\partial r} + \frac{\partial^2 v_r}{\partial r^2} + \frac{\partial^2 v_r}{\partial z^2} - \frac{v_r}{r^2}\right), \\ \frac{\partial v_z}{\partial t} + v_r \frac{\partial v_z}{\partial r} + v_z \frac{\partial v_z}{\partial z} - \frac{v_\theta^2}{r} = -\frac{1}{\rho}\frac{\partial \rho}{\partial r} + \nu\left(\frac{1}{r}\frac{\partial v_z}{\partial r} + \frac{\partial^2 v_z}{\partial r^2} + \frac{\partial^2 v_z}{\partial z^2} - \frac{v_z}{r^2}\right), \\ \frac{\partial v_\theta}{\partial t} + v_r \frac{\partial v_\theta}{\partial r} + v_z \frac{\partial v_\theta}{\partial z} - \frac{v_\theta v_r}{r} = 2\omega_0 v_r - \frac{1}{\rho}\frac{d\rho}{dr} + \nu\left(\frac{1}{r}\frac{\partial v_\theta}{\partial r} + \frac{\partial^2 v_\theta}{\partial r^2} + \frac{\partial^2 v_\theta}{\partial z^2} - \frac{v_\theta}{r^2}\right), \\ \frac{1}{r}\frac{\partial}{\partial r}(r v_r) + \frac{\partial v_z}{\partial z} = 0. \end{cases} \tag{8}$$

To solve Eq. (8), we will use the algorithm described in [7], according to which the solution procedure is divided into stages (Fig. 4).

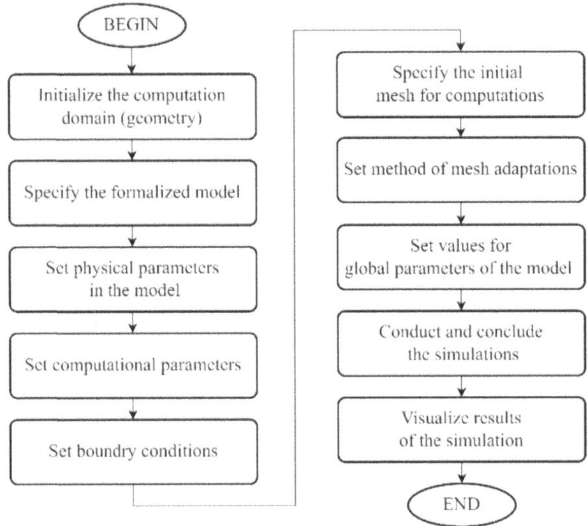

Fig. 4. Flow chart for setting the simulation in the FlowVision

The hydrodynamic modeling of the propeller-type micro HPP hydroturbine begins with the creation of the computational domain or geometry of the object in the CAD system Kompas-3D. This geometry is then imported into FlowVision as a document containing the computational domain, as shown in Fig. 5. The geometry represents a fluid volume that fills the propeller hydroturbine. The area illustrated in Fig. 5 is a fragment of the computational domain created for the modeled object in the FlowVision Preprocessor.

Within this domain, the FlowVision Preprocessor (the main module for preparing the computational project) defines the model as a fluid volume occupying the hydroturbine. In this domain, the following components are specified: the mathematical model, physical parameters, numerical method parameters, boundary conditions, and computational mesh. The calculation is then performed, meaning that this domain represents the completion of the preparatory stage before proceeding to simulation.

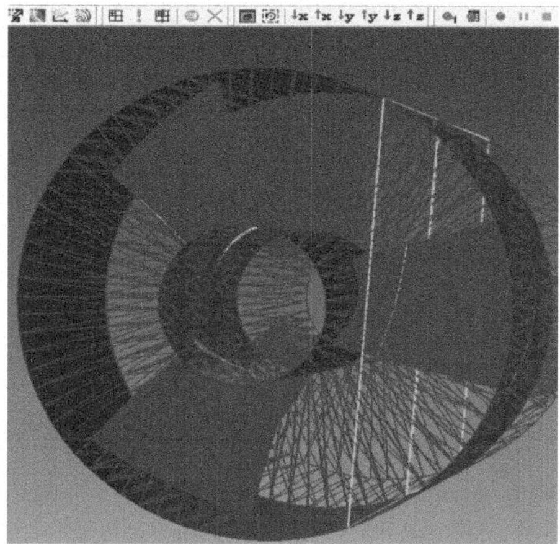

Fig. 5. Workspace with the object in FlowVision

The second stage of the simulation begins with defining the mathematical model itself.

FlowVision provides a wide range of basic models. For this case, the incompressible fluid model is selected, which is used to simulate fluid flow at high Reynolds numbers and low density variations.

The third stage involves specifying the physical parameters of the model, including the molecular mass of the fluid, density, and dynamic viscosity coefficient—that is, the properties of the medium in which the simulation is performed (in this case, water).

At the fourth stage, labeled "Method Parameters," the Preprocessor interface allows for the selection of the rotational motion type to simulate the rotation of the propeller turbine's runner induced by the energy of the water flow.

The fifth stage defines the boundary conditions, i.e., the physical constraints acting on the domain boundaries. Three boundary types are used in this model: fluid inlet to the runner, outlet from the runner, and walls or solid bodies that interact with the fluid (blades, hub, and the turbine housing walls).

In the sixth stage, the initial computational mesh is defined. FlowVision employs a local adaptive mesh approach, wherein the entire computational domain is divided into a specified number of intervals of fixed length.

The seventh stage configures mesh adaptation. This step improves the mesh resolution in geometrically complex or detailed regions of the model.

During the eighth stage, global parameters are set. This is done by accessing the "General Parameters" folder in the Preprocessor tree. It contains several configuration tabs; however, in this case, the default values are used.

The ninth stage consists of performing and completing the simulation.

Once the "Start" button is pressed, the simulation begins. The computation can be paused and resumed at any time to perform intermediate analysis or adjust parameters. Throughout the simulation, it is essential to monitor the iteration process to control numerical error. With correct setup, the computational error does not exceed 1% [7].

In the tenth stage, the simulation results are visualized.

This step is performed using the FlowVision Postprocessor. The results of the hydrodynamic interaction between the water flow and the turbine, as well as the hydrodynamic forces acting on the system, are shown in Fig. 6. The figure clearly illustrates the pressure distribution on the blade surfaces and the velocity streamlines. The "Layers" tab allows selection of various result display formats. The Postprocessor includes all the necessary tools for visualizing and integrating results over volumes, surfaces, and cross-sections. It also enables modification of simulation parameters and real-time visualization of intermediate results during the computation process.

Fig. 6. Simulation results of the hydrodynamics of a propeller-type hydroturbine

The FlowVision software suite incorporates standard tools required by designers during the development of technological systems. These standard tools are essential for analyzing the behavior of a system under a wide range of input parameters for the turbine runner, such as head H (m) and flow rate Q (m^3/s). The software enables the construction of performance characteristics critical to operational efficiency, with refinement of the parameters that influence these characteristics.

The main parameters of a micro HPP significantly affect the performance of the turbine runner design, and therefore must be adjustable in the simulation-laboratory test bench. This is because such a bench essentially functions as a physical modeling tool, which complements the mathematical modeling process. As a result, the outcomes of the computational model must be reproducible on the physical test bench, and the obtained data must be recorded using automatic recording or measuring instruments.

4 Simulation Results

When developing the simulation program, it is necessary to select a set of performance characteristics that determine the efficiency of the chosen turbine runner design. The most reliable assessment of system efficiency can be obtained by calculating and plotting the performance curve shown in Fig. 7, where the efficiency (CPE) of the hydroturbine serves as the primary indicator of the hydropower unit's effectiveness.

This characteristic makes it possible to determine that the considered turbine design can reach a maximum output power of 8 kW under specific head and flow rate values. Since this condition corresponds to the highest efficiency value, appropriate adjustments must be made to the micro HPP design to ensure optimal operation.

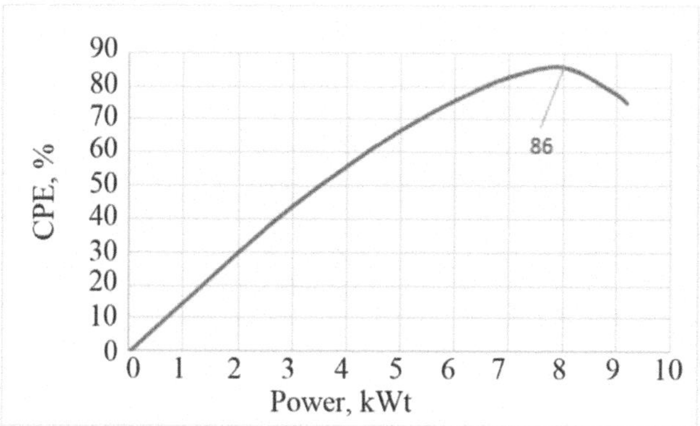

Fig. 7. Efficiency curve of the propeller-type hydroturbine

Figure 8 presents the operational characteristics for the considered turbine design, accounting for real operating conditions where maintaining a constant rotational speed is not feasible, as both head and water flow can vary significantly. Consequently, it is recommended to use operational characteristics derived from simulation data for different head values. In principle, the characteristics shown in Fig. 8 provide comprehensive information for the selected hydroturbine type and may prove useful when input parameters, such as the relationship between efficiency, power, and head, change, particularly when maintaining a constant rotational speed of the hydroturbine.

Based on the completeness of the information provided in the operational characteristics shown in Fig. 8, the characteristics of the runner wheel can be considered universal. Therefore, these characteristics can be used to achieve the objective outlined in the problem statement for this study, namely, to construct the mechanical characteristic that ensures the optimal operating regime of the hydroturbine.

Figure 9 shows the desired mechanical characteristic $\omega_{HT} = f\ (M_{HT})$ for the propeller-type hydroturbine, which was constructed following the sequence outlined below:

1. Data obtained from calculations using formulas (1–6) are entered into the FlowVision software package. Universal characteristics are constructed, indicating the regions

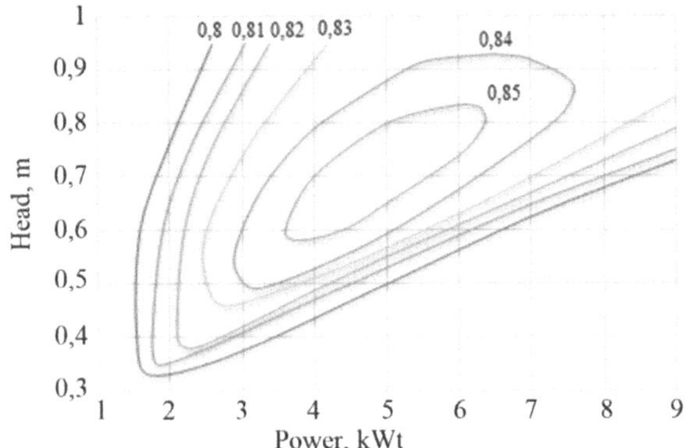

Fig. 8. Operational characteristic of the hydroturbine

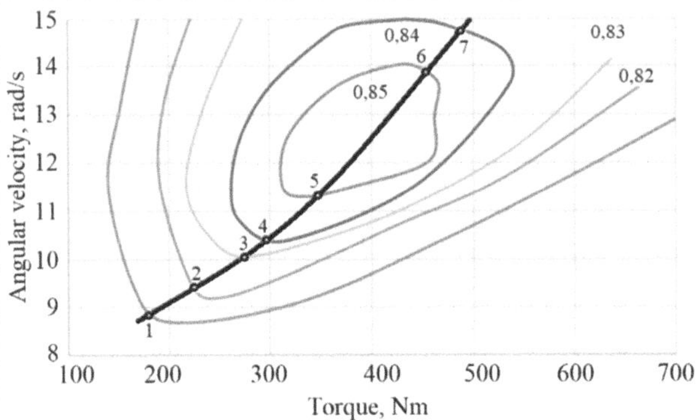

Fig. 9. Mechanical characteristics of the propeller-type hydroturbine $\omega_{\mathrm{HT}} = f\ (M_{\mathrm{HT}})$.

in which the selected runner wheel design develops turbine power, with efficiency values marked for a wide range of variations in the hydroturbine's input parameters. This approach is suitable for mountainous areas, where the turbine design under consideration will be operated.

2. The most likely operating zones are identified, where the turbine will operate under varying input parameters, and the optimal placement of the desired mechanical characteristic is determined from the perspective of efficient hydroturbine utilization, including zones where the efficiency value reaches 85%.
3. Considering that any structure driven into rotational motion by a directed flow, whether air or water, exhibits a fan characteristic, the points marked as 1, 2, ..., 7, passing through the breakpoints of the constraint regions, can be approximated and placed.
4. By drawing a curve through these points, guided by the fact that the load variation characteristic typical for centrifugal action devices is described by the fan characteristic formula, the desired curve is constructed, which is referred to as the mechanical

characteristic of the propeller-type hydroturbine [9]. In this case, the developed turbine power will correspond to the efficiency values determined by the zone where the curve $\omega_{\mathrm{HT}} = f\ (M_{\mathrm{HT}})$ lies, under varying input parameters of the runner wheel.

5. Although the mechanical characteristic, highlighted in black in Fig. 9, is constructed approximately, it can be considered as an interpreted result in the first approximation. This allows it to be represented mathematically by approximation or in graphical or tabular form, as a numerical representation. This significantly simplifies the inclusion of the characteristic in the required scale in the nonlinear block shown in Fig. 1.

5 Rationales for Control Loop Design

Figure 10 shows the essentials of schematic layout of the simulation laboratory test stand, which takes into account the specifics of selecting and constructing all elements, as detailed in this article.

In this diagram, only the simulation of the synchronous generator is described, which represents an automatic voltage regulation system based on the deviation of the root mean square value of the generator's output voltage [10].

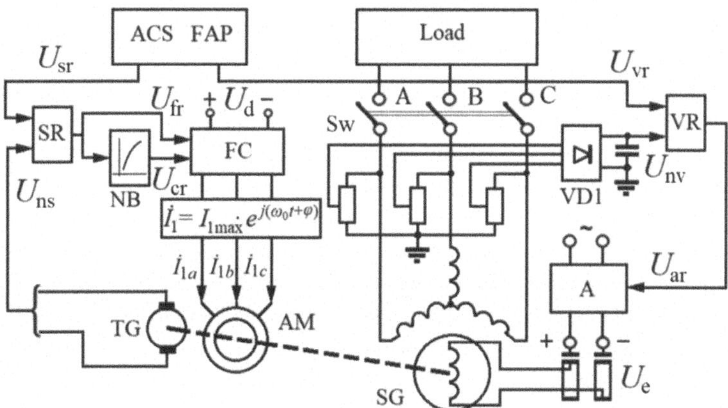

Fig. 10. Simplified schematic layout of the proposed physical similarity laboratory stand

To regulate the output voltage of the synchronous generator, an amplifier (denoted as A) is used. This amplifier supplies rectified network voltage to the excitation winding as a controllable excitation voltage U_e. The input signal to the amplifier is U_{ar}, which is formed at the output of the voltage regulator VR. The VR operates based on deviation. The input signals to VR are U_{vr} from the ACS FAP (Automated Control System for Frequency and Active Power) block, and the feedback signal U_{nv}. Voltage U_{nv} is proportional to the output voltage of the SG and is obtained from the rectifier VD1, which is assembled in a bridge circuit with subsequent filtering.

The generated electrical power from the stator windings of the generator is supplied to the load through a switch. The load during experiments can consist of three-phase circuits containing resistive, capacitive, and inductive elements. The last type of load is a standard three-phase asynchronous motor, which allows for load variation within a certain range.

The laboratory simulation stand includes a computing system that allows simulating the automation of electricity generation for micro HPP, such as the 'Automated Control System for Frequency and Active Power' mode, as shown in Fig. 10. It should be noted that the micro HPP operation program is defined by the ACS FAP, which sets the operating point on the mechanical characteristic $\omega_{HT} = f\ (M_{HT})$, followed by the stabilization of the micro HPP parameters.

The setting of the operating point for the micro HPP is carried out using two coordinates: the speed ω_{HT}, set by the speed regulator SR based on the signal U_{fr}; and the current associated with the torque of the hydro turbine, set by the nonlinear block NB using the signal U_{cr}. In this process, the frequency converter FC, operating in current source mode, generates a three-phase system of symmetrical currents $\dot{I}_{1a}$, $\dot{I}_{1b}$, $\dot{I}_{1c}$ to induction asynchronous motor AM, with adjustable parameters for amplitude $I_{1\max}$, frequency ω, and phase φ, following a law described by the following equation [5]:

$$\dot{I}_1 = I_{1\max} \cdot e^{j(\omega_0 t + \varphi)}. \quad (9)$$

6 Results

By analyzing the characteristics presented in Fig. 9, several conclusions can be drawn. The FlowVision software package enables the construction of universal characteristics that highlight regions where the selected impeller design operates with turbine power, which is developed at specified efficiency values. This approach allows for a comprehensive understanding of the performance of the turbine under various conditions.

From the perspective of optimizing the hydro turbine's efficiency, it is crucial to begin constructing the mechanical characteristic in the region where the efficiency reaches 85%. This threshold is essential for ensuring the effective operation of the turbine. The desired mechanical characteristic curve, $\omega_{HT} = f\ (M_{HT})$, should encompass a wide range of input parameters and pass through the breakpoints of the constraint regions, indicating where the turbine is likely to operate most efficiently.

The mechanical characteristic, highlighted in black in Fig. 9 for the selected turbine design, can be considered the operational characteristic. To facilitate further analysis, it should be approximated and expressed as a mathematical function. If this is not feasible, the characteristic can be represented numerically, either in graphical or tabular form, which significantly simplifies its integration into the nonlinear block.

References

1. Kurmanov, N., Kabdullina, G., Baidakov, A., Kabdolla, A.: Renewable energy, green economic growth and food security in central asian countries: an empirical analysis. Int. J. Energy Econ. Policy **15**(2), 1–8 (2025). https://doi.org/10.32479/ijeep.17922
2. Laldjebaev, M., Isaev, R., Saukhimov, A.: Renewable energy in Central Asia: an overview of potentials, deployment, outlook, and barriers. Energy Rep. **7**, 3125–3136 (2021). https://doi.org/10.1016/j.egyr.2021.05.014

3. Azimov, U., Avezova, N.: Sustainable small-scale hydropower solutions in Central Asian countries for local and cross-border energy/water supply. Renew. Sustain. Energy Rev. **167**, 112726 (2022). https://doi.org/10.1016/j.rser.2022.112726
4. Rodríguez, L., Sanchez, T.: Designing and building mini and micro hydro power schemes. Practical Action Publishing, Rugby, UK (2011)
5. Veltman, A., Pulle, D.W.J., De Doncker, R.W.: Fundamentals of electrical drives. 2nd edn. Springer, Cham (2016). https://doi.org/10.1007/978-3-319-29409-4
6. Tan, K.K., Putra, A.S.: Drives and control for industrial automation. Springer, Cham (2011). https://doi.org/10.1007/978-1-84882-425-6
7. Mederov, T.T.: Research and Development of a Birotor Microhydroelectric Power Plant. Dissertation for the degree of Candidate of Technical Sciences, defended March 3 2017, approved June 23 2017, Kyrgyz State Technical University, Bishkek (2017). (in Russian)
8. Hailu, G., Varchola, M., Hlbocan, P.: Design of hydrodynamic machines pumps and hydro-turbines. Routledge, Milton Park, UK (2022)
9. Aksenov, A.A.: FlowVision: Industrial computational fluid dynamics. Comput. Res. Model. **9**(1), 5–20 (2017). https://doi.org/10.20537/2076-7633-2017-9-5-20
10. Kadyrov, I., Karaeva, N., Andarbekov Zh., Bakyt uulu, A., Fedorov O., Vladimirov O. Automatic voltage regulation system construction for synchronous generator of a small hydro power plant using thyristor pathogens. E3S Web Conf. **178**, 01036 (2020). https://doi.org/10.1051/e3sconf/202017801036

Computing Technologies in Data Analysis and Decision Making

Computer Modeling of the Surface of Unwrapped Fabric Using Computer Tools for Production Automation

Mikhail Agafonov(✉) and Igor Kondakov

Reshetnev Siberian State University of Science and Technology, Krasnoyarsk, Russia
mihail.agafonov.01@mail.ru

Abstract. Modeling of the fabric surface is the difficult task in terms of definition of the surface shape. Automating the modeling process as part of the textile production process can reduce development time for textile systems, such as fabric weaving machines that use the technique of cutting woven hoses. In this paper, a method is proposed for computer modeling of the surface of fabric unwrapping from a cylindrical shape to a flat one without stretching. The proposed method consists of next five stages: determining the pattern of changes in the surface cross-section, development of the virtual instrument in the graphical programming environment "LabVIEW", creating the point cloud and using it in CAD system "KOMPAS-3D". The virtual instrument makes sequential changes to the parameters of the surface cross-section, creates the point cloud of the surface and exports the data to a file for further processing. The file is imported into the CAD system where the surface is building based on the point cloud data.

Keywords: Production Automation · Computer Technology · Surface Modeling · Computer Modeling · Textile Industry · Aerospace Industry · Carbon Fiber

1 Introduction

Currently, the aircraft, aerospace, and textile industries are in the process of development and improvement of their manufacturing capabilities throughout all stages of production. Known technologies for the production of composite materials are being developed towards automation. For instance, there is significant discussion regarding the automatic production of flat carbon fiber fabric with a predetermined pattern. The use of carbon fiber fabric, which is woven or twisted from several carbon tapes, allows create a lighter structure while maintaining the strength properties. This could potentially influence the final cost of a product, such as the cost of launching one kilogram of payload into near-Earth orbit.

For the production of flat fabric can be used a technique of weaving a cylindrical preform with fiber. This technique allows to create closed hoses, which is cut to obtain a flat shape. In order to design a system that works with on technique, it is essential to the shape of the resulting fabric. A precise description of the finished surface and its

V. Jordan et al. (Eds.): HPCST 2024, CCIS 2919, pp. 103–115, 2026.
https://doi.org/10.1007/978-3-032-20325-0_8

3D modeling allows to utilize the resulting 3D model for further modeling of complex production systems.

2 Problem Statement and Proposed Solution

There are various options for the manufacturing of flat fabrics using automatic weaving machinery, such as a braiding machine [1], which weaves a tubular core with flat yarns to produce fabric hoses. It is possible to develop on the basis of this braiding machine a device capable of transforming a tubular fabric into a flat sheet, as described in the patent [2]. The production process using this method involves several stages. First, the required number of threads with the specified properties are charged into a circular weaving machine. Subsequently, a cylindrical workpiece, which has been pre-installed on the weaving machine is woven. Next, in order to obtain a flat fabric, a device must be installed for cutting and straightening the fabric, transforming it into the desired form.

In order to model the structure, it is necessary to visualize the shape of the surface that the fabric assumes during its unwrapping process. There are several options for representing the shape of the surface: analytical, manual (through surface modeling), or a combination of the two. According to the author, the most straightforward and accurate method is the combined approach, which utilizes a graphical programming environment "LabView" [3] and CAD system "KOMPAS-3D" [4].

To begin working on shaping the desired surface, it is necessary to establish the conditions for fabric transformation. Carbon fiber fabric has a low-tension coefficient, so it can be assumed to be inextensible. Additionally, the desired fabric shape should not have any folds or irregularities on the surface or self-intersections.

For instance, manual modeling of the surface of a fabric unwrapping from tubular to flat allows to get a general view of the fabric, as shown in Fig. 1 (hatched for clarity). This instance was built by creating a surface using sections that were manually created in a CAD system "KOMPAS-3D". During the modeling process, several drawbacks of this method have been identified. The overall difficulty of the modeling increases, as does the need for greater accuracy of the final surface. Additionally, there are issues with the undulations of the fabric at the edges, as well as control over the width of the fabric and its stretching.

Therefore, in order to accelerate the modeling process and achieve a more accurate result, it is essential to utilize automated modeling methods. These methods should include the automatic calculation of points of the fabric surface that will be used to create a 3D model of the fabric.

3 Modeling of Fabric Surface Using Combined Method

For the combined method of fabric modeling, it is necessary to determine the pattern of change in the shape of the section. To do so, create a sketch indicating all relevant values. In order to gradually transform the fabric from a cylindrical to a flat shape, it is necessary to build several sections at a specified distance from one another. Further, in each subsequent section, the diameter of the circle applied to the fixed point should be progressively increased, and an arc segment of the same length deposited on it. Then,

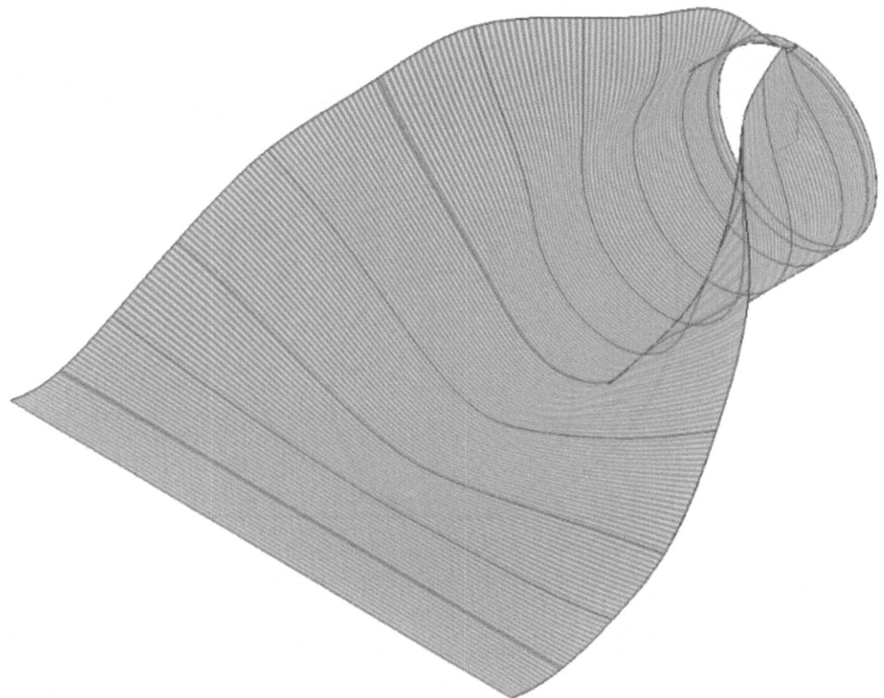

Fig. 1. The unwrapped surface of the fabric created by manual modeling method

with an infinitely increasing diameter, the drawn segment of a fixed length will become straight, and the surface formed according to the resulting sections will be the desired surface for the unwrapped fabric. A graphical representation of this method is shown in Fig. 2.

The principle of sketching may be summarized as follows:

$$L_1 = \pi D_1, \tag{1}$$

$$L_2 = \frac{\pi D_2 \alpha}{360}, \tag{2}$$

$$\alpha = \frac{360 D_1}{D_2}, \tag{3}$$

$$L_n = \frac{\pi D_n \alpha}{360}, D_n \to \infty \tag{4}$$

where D_i – the diameter of the circular section at the ith instance of time, α – the angle of the circular section at the ith instance of time.

Since the fabric unwrapping process is symmetrical, which simplifies calculations and production processes, it is only necessary to create half of the surface on one side of the vertical axis and then mirror the other half. In the graphical programming environment "LabVIEW", the virtual instrument has been developed that performs these

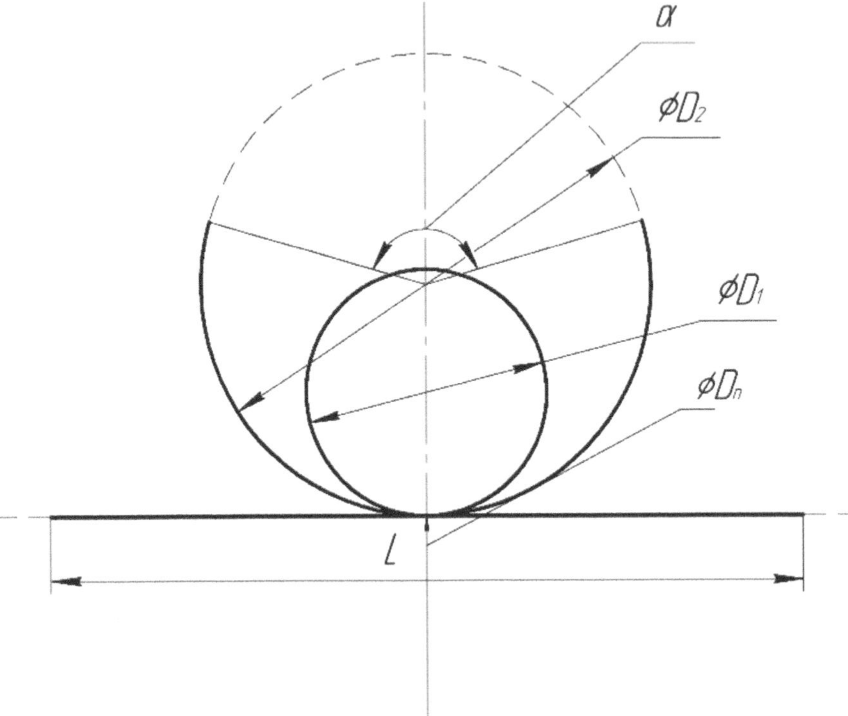

Fig. 2. A sketch of sections of the unwrapping fabric

transformation operations. The inner loop of the program creates an arc of the polygon based on the specified parameters and also displays data on a graph. The external loop initializes the creation of the table [5], comprising three columns (x, y, and z), which correspond to the coordinates of a set of surface points. Additionally, within the external loop, there is a progressive increase in the radius of a circle and a parameter (z) that determines the separation between sections. The block diagram of the virtual instrument is shown in Fig. 3.

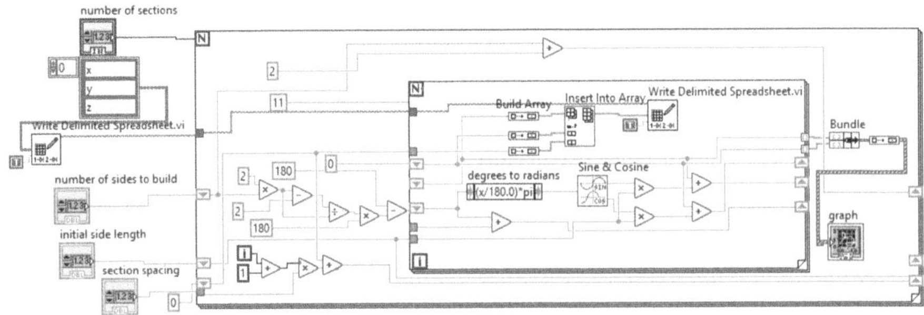

Fig. 3. The block diagram of the virtual instrument

On the front panel of the virtual instrument, which is shown in Fig. 4, users can configure parameters such as the number of sections, number of points per section, initial side length, and the distance between the sections. Each surface section is displayed on a graph to visualize the resulting shape.

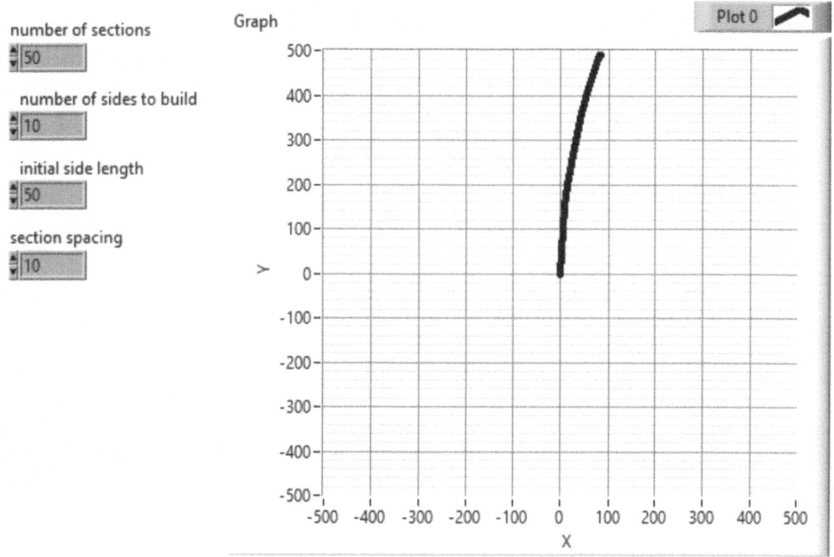

Fig. 4. The front panel of the virtual instrument

After starting the virtual device, the user has to select a location to save the generated point table and also specify the resolution of the resulting.xls file. This generated table, as shown in Fig. 5, is essential for further construction with the CAD system "KOMPAS-3D".

It is possible to automate the process of finding various forms of the unwrapped surface. To accomplish this, the virtual instrument must be supplemented with a fragment of the point cloud output in 3D format. To implement this function, the source virtual instrument must be extended with functions that generate a 3D environment and initialize a sphere with a large diameter, which will serve as the background of the required color. Subsequently, the individual points of the intended surface are added to the scene as a geometric mesh. Following this, the display colors and camera position in space are configured.

External variables can be used to aid in debugging the program and to modify the shape of the surface. They can also be used to organize parameterized surface modeling, depending on the needs of production. The block diagram and the front panel of the virtual instrument are shown in Figs. 6 and 7, respectively.

To create 3D model of the surface, the user should utilize the CAD system "KOMPAS-3D" and call the "Surface by Point Layer" command from the "Frames and

	A	B	C
1	x	y	z
2	0	0	0
3	15,451	47,553	0
4	44,84	88,004	0
5	85,291	117,393	0
6	132,844	132,844	0
7	182,844	132,844	0
8	230,397	117,393	0
9	270,847	88,004	0
10	300,237	47,553	0
11	315,688	0	0
12	315,688	-50	0
13	0	0	10
14	12,941	48,296	10
15	37,941	91,598	10
16	73,296	126,953	10
17	116,598	151,953	10
18	164,894	164,894	10
19	214,894	164,894	10
20	263,19	151,953	10
21	306,491	126,953	10
22	341,847	91,598	10
23	366,847	48,296	10
24	0	0	30
25	11,126	48,746	30
26	32,82	93,795	30
27	63,995	132,886	30
28	103,086	164,061	30
29	148,135	185,755	30
30	196,881	196,881	30
31	246,881	196,881	30
32	295,628	185,755	30
33	340,676	164,061	30
34	379,768	132,886	30
35	0	0	60
36	9,755	49,039	60

Fig. 5. The fragment of the generated point cloud table

Surfaces" section [6, 7]. Subsequently, a file with the point cloud should be selected and the modeling parameters are specified: the type of surface generation is pole spline, and the method of generation is automatic. As a result, the required 3D model is generated in a polygonal form, which is shown in Fig. 8.

Using the resulting surface, a solid-state model with the desired thickness can be created. To do so, the "add thickness" command must be called. The resulting solid-state model is shown in Fig. 9. This model may be utilized in the further development of a textile system or for fabric analysis within CAE) systems.

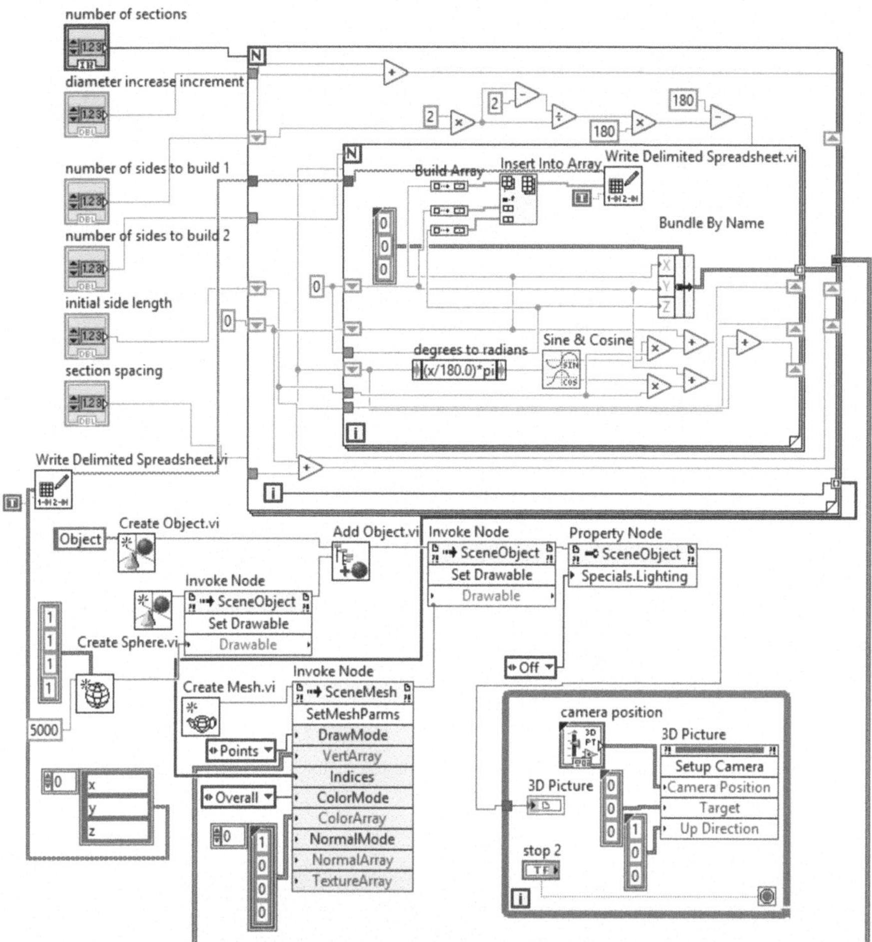

Fig. 6. The block diagram of the extended virtual instrument

Therefore, the virtual instrument has been developed that consists of two main components: a component for determining the parameters of the functions to be processed and a component for displaying points based on the performance of those functions. The virtual instrument is capable of generating a point cloud in accordance with a given law of sectional change, even if that law of change were to be altered. For instance, one could specify the shape of a surface transformation from parabolic to semicircular or other variations that could be utilized in the manufacturing of equipment for textile production or other purposes. It is likewise possible to modify the displacement pattern of each successive section. This could be employed to simulate the transformation of a fabric surface along a complicated trajectory.

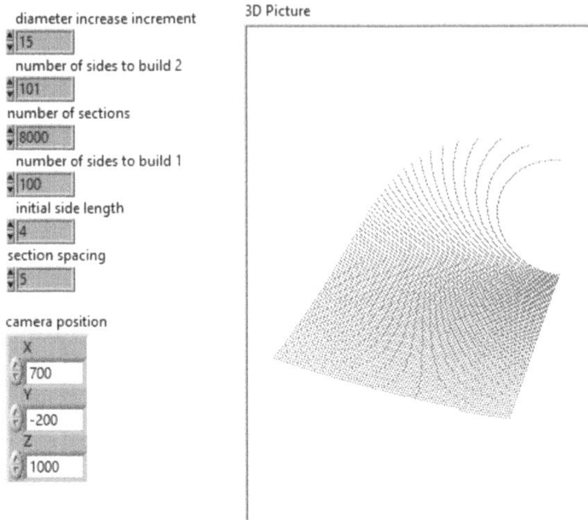

Fig. 7. The front panel of the extended virtual instrument

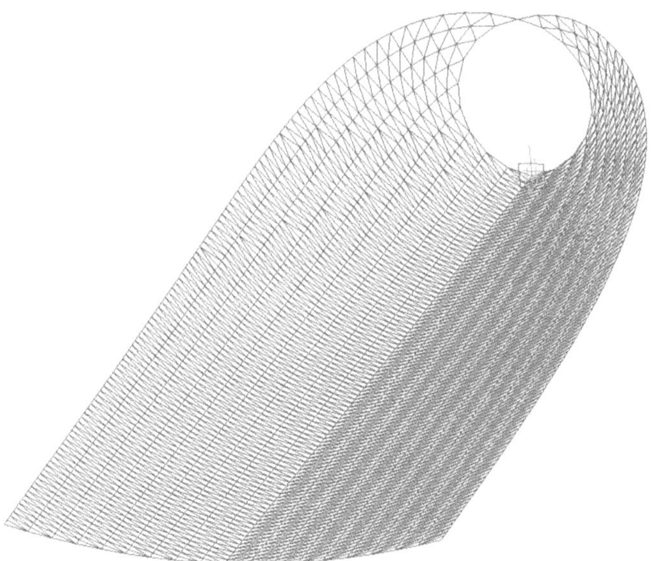

Fig. 8. The generated 3D model of the surface in a polygonal form

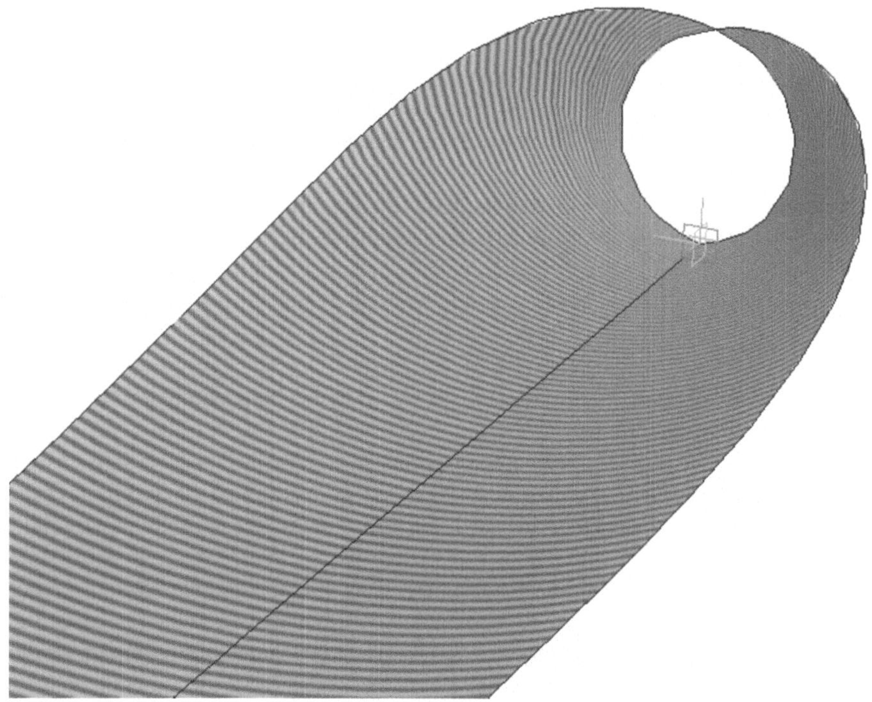

Fig. 9. The solid-state fabric model

4 Additions to the Modification of Virtual Tool

If the task is to model the surface of a fabric transformation along a complex path, the first step is to establish certain conditions that restrict the mathematical processing of the virtual instrument. The path along which the segments will move must be defined as a curved line formed by a set of points with large radii of curvature. The segments should be positioned at a sufficient distance from each other to prevent intersections or overlapping of the points of two adjacent segments. The segments should be perpendicular to the tangent of the curve at the point where they are generated.

The curved line can be generated from a set of small line segments, or it can be smoothed programmatically. A smoothed version of the curve can be a Bezier curve, which simplifies the process of creating a curved line but may reduce the precision of the final model. One method for constructing Bezier curves, described in [8], involves defining several points along the desired path of the curved line. This method is based on using these points as control points for the Bezier curve.

After the path of the segments has been determined, it is necessary to create the segments at predetermined locations. If the path is defined by a series of cubic Bezier curves, then the tangent line at a given point can be calculated by differentiating the specific curve according to its known equation.

The equation of the cubic Bezier curve is given below:

$$S_0(t) = (1-t)^3 \cdot P_0 + 3 \cdot t \cdot (1-t)^2 \cdot P_1 + 3 \cdot t^2 \cdot (1-t) \cdot P_2 + t^3 \cdot P3 \quad (5)$$

where P_i – control points for which the curve is defined, t – the argument responsible for moving along the curve.

The tangent to the Bezier curve can be defined as the derivative of the curve at a specific point. This can be expressed mathematically using the following equation:

$$S_0(t)' = -3P_0(1-t)^2 + 3P_1 \cdot (1-t)^2 - 6t \cdot P_1(1-t) - \\ - 3P_2 + 6t \cdot P_2(1-t) + 3 \cdot P_3 t^2 \quad (6)$$

Next, at a given point, a normal plane is created that is perpendicular to the tangent line to the curve at that point.

In order to convert the coordinates of points of a generated section into absolute coordinates, it is necessary to utilize an auxiliary virtual instrument that multiplies transformation and displacement matrices with the coordinates of each individual point in the constructed section.

A mathematical coordinate transformation using rotation and translation matrices is known [9]. The method involves finding the coordinates and angles by which the initial coordinate system must be rotated and translated so that it aligns with the current system. To accomplish this, at a specified point, the angles of incidence of the tangents to the Bezier curves and the distances along the X, Y, and Z axes from the origin must be determined. Subsequently, the coordinates of the intersection points recorded using a matrix are multiplied by a transition matrix. The block diagram of the virtual instrument that implements this method is illustrated in Fig. 10.

If necessary, it may be possible to streamline the process of fabric unwrapping by supplementing the virtual instrument with a mathematical algorithm for converting an arbitrary given curve into a planar one while preserving its length. There is a well-known method for transforming a curved line into a planar one, as described in reference [10]. In order to extend an arbitrary line, it is necessary to define the initial segment using a point cloud. These points can be determined using the previously developed program. An example of defining a point cloud is illustrated in Fig. 11. Subsequently, a curved polyline must be created by connecting the defined points with line segments. To reverse the polyline, a direct kinematic problem must be solved in order to modify the angles between adjacent line segments such that the overall curve tends towards a straight line. Given the fixed lengths of the line segments, the fabric width will remain constant at all times, as required by the conditions for maintaining the fabric's shape.

Therefore, an enhanced version of the program has been developed for generating a curved surface unwrapping. This version offers the advantage of being able to generate a point cloud representation of any complex curved surface. However, the accuracy of the unwrapped surface may be slightly lower.

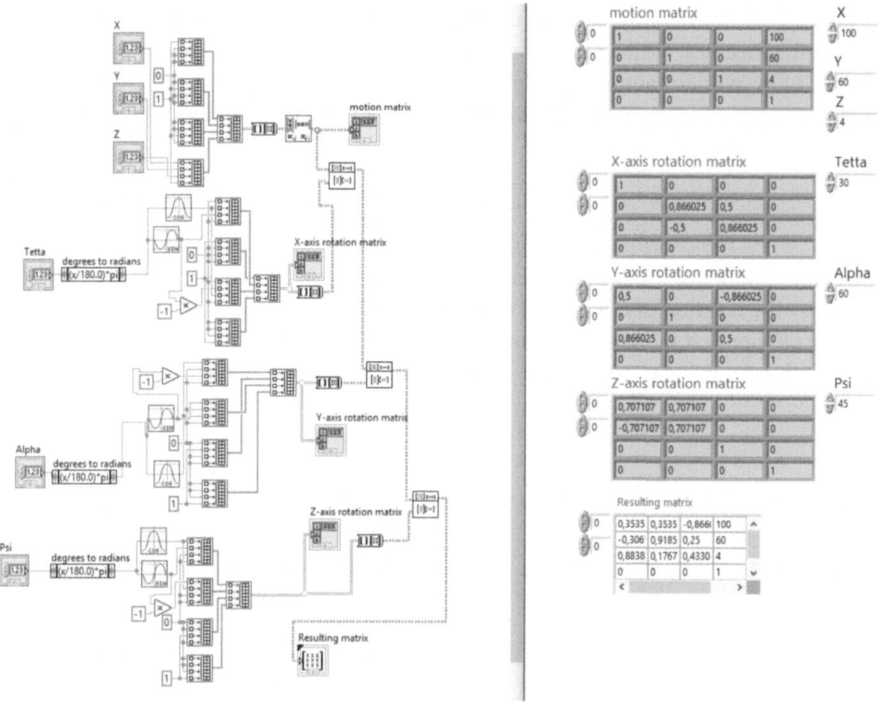

Fig. 10. The block diagram of the virtual instrument to coordinate transform

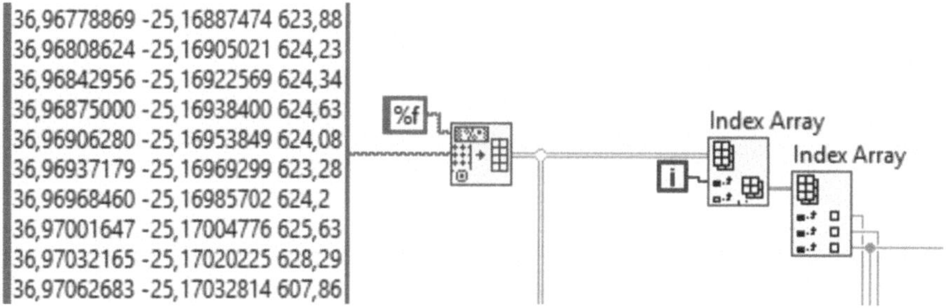

Fig. 11. A fragment of demonstrating the use of a variable to set point cloud

5 Testing the Program in Equipment Development

Another potential application for the virtual instrument would be the development of a surface for the technology of layering of carbon fiber. The layering process involves impregnating carbon fiber with an adhesive and laying it out on specialized equipment, either joint-to-joint or with some degree of overlap. After laying out one-layer, additional layers can be applied, but at a different angle to the original layer. The benefit of this process is the creation of a composite material with superior strength characteristics [11]. The process can be carried out manually or using specialized robotic systems.

One of the most promising applications for this process is in the production of spacecraft reflectors. Reflectors produced in this manner have a very low weight-to-strength ratio, while maintaining high structural integrity. In general, the shape of the antenna's surface is described by the equation for an elliptical parabolic shape:

$$z = \frac{x^2}{a^2} + \frac{y^2}{b^2} \tag{7}$$

where x, y, z – the coordinates of a three-dimensional graph, a and b – arguments that affect the curvature of the graph. For a = b, an elliptical paraboloid is a surface of rotation formed by the rotation of a parabola around its axis of symmetry.

Considering that the laying out of the material should be carried out on an inverted template, it is possible to modify the equation for the desired surface as follows:

$$0 < z < c - \left(\frac{x^2}{a^2} + \frac{y^2}{b^2}\right), \tag{8}$$

where c – the argument responsible for the total height of the resulting surface.

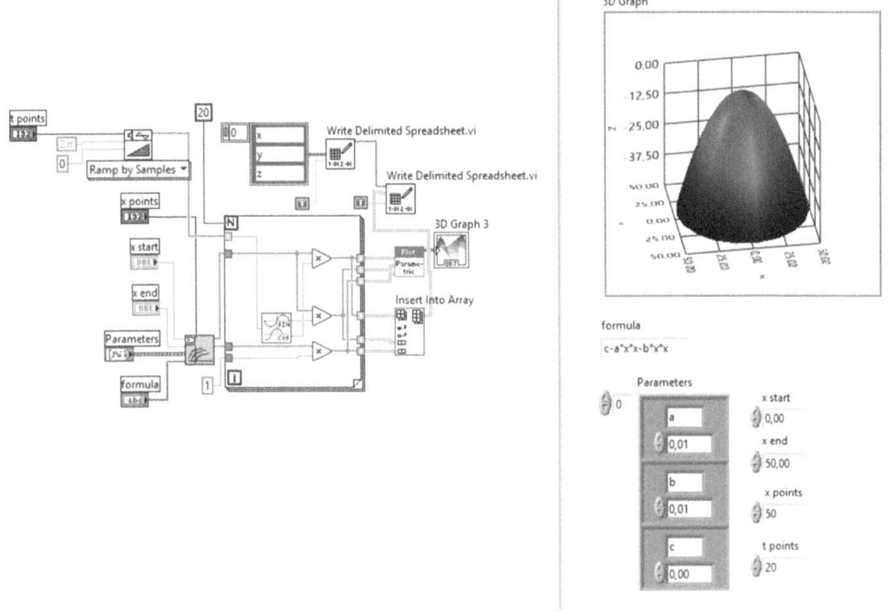

Fig. 12. The virtual device for the development of a surface for the technology of layering of carbon fiber.

To create the desired surface, a virtual instrument can be created similar to the previously described ones. Its purpose is to establish a formula and defines the parameters for generation of the surface. Subsequently, it is necessary to perform the initialization process of the table, during which the names of the columns are determined. The parameters of the function arguments are set by external variables, which are then processed in the main program loop. Following mathematical processing, data regarding each point is presented on a 3D graph and a surface is generated for a more visually appealing display. The diagram block and the front panel of the virtual device performing the assigned tasks are shown in Fig. 12.

Therefore, a virtual instrument has been developed that generates a surface of elliptical paraboloid that is symmetrical about the Z-axis. This virtual device facilitates the generation of numerous curved surfaces that are symmetrical about to the Z-axis. To create surfaces of other shapes, a user must input a formula describing the desired shape in the relevant field on the front panel of the virtual device.

6 Conclusion

As a result, the proposed method for constructing complex surfaces has the potential to be applied in the development of devices that interact directly with fabric and modify them for production purposes. Additionally, the construction algorithm could be modified to create other shapes necessary for the design process. By combining software packages, it is possible to obtain a parameterized model that is loaded from an external file, thereby speeding up the overall time required for the development of complex surfaces and improving the quality of the design.

References

1. A medical device including a tubular structure and a method for making the same : pat. 2699719 Europe. № 12774865.5 ; fil. 17.04.12 ; publ. 06.03.19, Bull. 2019/10. 30 p.
2. Warp beam for triaxial weaving : pat. 3,884,429 United States. № 395,529 ; fil. 10.09.73. ; publ. 20.05.75. 10 p.
3. Kolker, A.B., Livenets, D.A., Kosheleva, A.I.: Justification of the choice of software for robotics. Automation and software engineering **1**(1), 51–64 (2012)
4. Chernousov, E.A.: Computer-aided design systems in the implementation of aircraft engineering tasks. Step into Science **4**, 81–86 (2022)
5. Korgin, A.V., Emelianov, M.V., Ermakov, V.A., Zeid Kilani, L.Z., Krasochkin, A.G., Romanets, V.A.: The use of LabVIEW to solve the problems of collecting and processing measurement data in the development of monitoring systems for load-bearing structures. Vestnik MGSU **9**, 135–142 (2013)
6. Terin, A.M., Tutushkin, A.K., Pankov, D.E., Solomonov, I.A.: Complex surface modeling in CAD system. Current problems of aviation and cosmonautics **1**, 46–48 (2020)
7. Kaminskiy, A.I.: Three-dimensional modeling in the study of engineering graphics. Academy **4**(31), 10–11 (2018)
8. Faraway, J.J., Reed, M.P., Wang, J.: Modelling Three-Dimensional Trajectories by Using Bézier Curves with Application to Hand Motion. J. R. Stat. Soc.: Ser. C: Appl. Stat. **56**(5), 571–585 (2007). https://doi.org/10.1111/j.1467-9876.2007.00592.x
9. Fu K.S., Gonzalez R.C., Lee C.S.G. Robotics: control, sensing, vision and intelligence: McGraw-Hill, 624 (1987)
10. Su, Z., Li, L., Zhou, X.: Arc-length preserving curve deformation based on subdivision. J. Comput. Appl. Math. **195**(1–2), 172–181 (2006)
11. Fan, Q., Duan, H., Xing, X.: A review of composite materials for enhancing support, flexibility and strength in exercise. Alex. Eng. J. **94**, 90–103 (2024)

Algorithms for Recognition that Rely on the Creation of One-Dimensional Threshold Rules

Gulmira Mirzaeva[1,2], Movludaxon Nugmanova[1], Ganidjan Porsayev[3], Ravshan Shirov[3], and Nomaz Mirzaev[1,2](✉)

[1] Research Institute for the , Development of Digital Technologies and Artificial Intelligence, 17A, Boz-2, Tashkent, Uzbekistan 100125
nomazmirzaevich@gmail.com

[2] Tashkent University of Information Technologies Named After Muhammad Al-Khwarizmi, 108, Amir Temur Avenue, Tashkent, Uzbekistan 100084

[3] Samarkand State University Named After Sharof Rashidov, 15, University Blv., Samarkand City, Uzbekistan 140104

Abstract. This article explores the creation of recognition algorithms (RAs) utilizing one-dimensional threshold rules (OTRs) to address the challenge of classifying objects within the complex high-dimensional feature space (HDFS). A novel method is introduced, centered around the establishment of a collection of basic objects (BOs) and the subsequent development of an OTR for each. What sets this approach apart is the construction of OTRs derived from each chosen base object. The proposed RA is defined parametrically through a series of sequential steps designed to address specific subtasks. These key procedures include: determining the difference function (DF) in the subspace of representative features (SRF); identification of subsets of interrelated objects (SIOs); formation of a set of BOs; construction of tabular proximity functions (TPFs); calculating the proximity score between (PSB) the base and simple objects; calculating the PSB a class and an object. To evaluate how well the suggested RAs model works, we conducted experiments to address object classification, a model problem. The results of comparing the suggested RAs with existing ones are included.

Keywords: Recognition Algorithm · Representative Feature · Difference Function · Group Of Interrelated Objects · Base Object · Tabular Proximity Function

1 Introduction

Investigation of existing publications, in particular [1–5], It demonstrates that in the early days of pattern recognition development, these methods were primarily used to tackle problems with poorly defined fields like medical diagnosis, geological prediction, and sociological process classification. Research conducted during this initial period produced numerous recognitional algorithms (RAs) designed for specific practical applications. The research conducted during this initial period resulted in the emergence

V. Jordan et al. (Eds.): HPCST 2024, CCIS 2919, pp. 116–127, 2026.
https://doi.org/10.1007/978-3-032-20325-0_9

of numerous recognition algorithms (RA) designed for specific practical applications. They were based mainly on heuristic principles for assessing the proximity of recognized objects to given classes [1, 4]. The assessment of their performance was determined by the experimental results achieved [5, 6].

Subsequently, on the basis of the most common heuristic RAs, families of such algorithms began to be formed, united by a single structural description and parametric representation. This made it feasible to formulate and solve the problem of constructing an extremal algorithm within the framework of a certain RAs model. In Fig. 1 shows the main models of RAS. To date, these models have been studied quite deeply:

R-models [7–12], S-models [9, 12–15], P-models [9, 16–21], L-models [22–27], O-models [1, 2, 28–33].

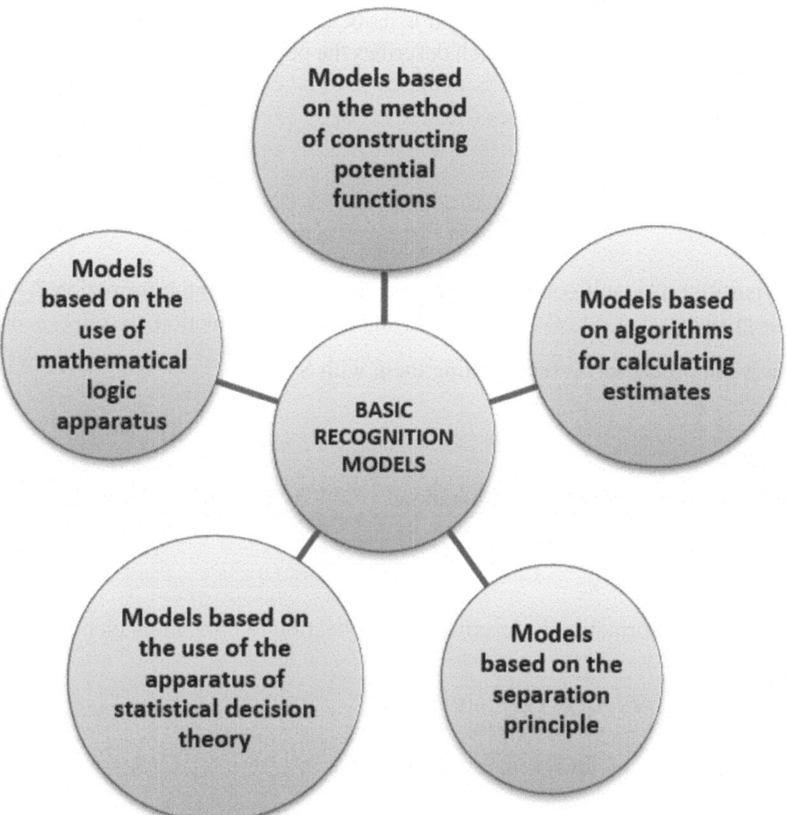

Fig. 1. Classification of basic recognition models

As shown in [5, 6, 34], these RAs are mainly focused on solving problems and recognition under conditions of independence (or weak dependence) of signs. In last years, various applied problems related to the recognition of objects described in the HDFS have often arisen. Acknowledged [34–36], that in many problems where objects are described in HDFS, the assumption of independence of features may not be satisfied. This suggests that the practical suitability of specific RAs for tasks involving HDFS object analysis is an under-explored area. Thus, the classification of objects characterized in HDFS is a significant challenge. This study aims to create RAs by constructing OTRs for addressing object recognition problems with in HDFS. To tackle these issues, we propose a method centered on building BOs within the RFs space and generating TPFs from these BOs. Note that the terminology and symbols used are adapted from [1, 4, 30].

2 Basic Concepts and Notation

Consider a set of valid objects, denoted by $\mathbb{X}$, defined within an n-dimensional feature space $\mathcal{X}$. Each valid object $S(S \in \mathbb{X})$ in is associated with a numerical representation (an n-dimensional vector) $\mathcal{I}(\mathrm{S})$, which describes the object. This is expressed as $\mathcal{I}(\mathrm{S}) = (x_1, .., x_u, .., x_n)$ [1, 4]. The set $\mathbb{X}$ is understood to be composed of ℓ non-overlapping subsets, often referred to as classes $\mathcal{C}_1, .., \mathcal{C}_{\mathrm{j}}, .., \mathcal{C}_\ell$:

$$\mathbb{X} = \bigcup_{\mathrm{j}=1}^{\ell} \mathcal{C}_{\mathrm{j}}, \mathcal{C}_{\mathrm{i}} \cap \mathcal{C}_{\mathrm{j}} \neq \varnothing,\ \mathrm{i} \neq \mathrm{j},\ \mathrm{i}, \mathrm{j} \in \{1, \ldots, \ell\}. \tag{1}$$

The partition (1) isn't completely defined; just some starting information $\mathcal{I}_0$ on the classesis given: $\mathcal{C}_1, .., \mathcal{C}_{\mathrm{j}}, .., \mathcal{C}_\ell$. To illustrate this starting information, we select m objects from the valid object set $\mathbb{X}$, representing them with $\overset{\sim \mathrm{m}}{\mathrm{S}}$:

$$\overset{\sim \mathrm{m}}{\mathrm{S}} = \{\mathrm{S}_1, \ldots, S_{\mathrm{u}}, \ldots, S_{\mathrm{m}}\}.$$

Now, let's define the following notation:

$$\tilde{\mathcal{C}}_j = \overset{\sim \mathrm{m}}{\mathrm{S}} \cap \mathcal{C}_{\mathrm{j}}, \tilde{\mathfrak{D}}_{\mathrm{j}} = \overset{\sim \mathrm{m}}{\mathrm{S}} \backslash \tilde{\mathcal{C}}_j$$

The starting information is described as a collection $\mathbb{I}_0$ containing elements that are pairs:

$< S_i, \alpha(S_i) > (\forall S_i \in \mathbb{X}, i = \overline{1, m})$:

$$\mathcal{I}_0 = \{< S_1, \beth(S_1) >, \ldots, < S_i, \beth(S) >, \ldots, < S_m, \beth(S_m) >\}, \tag{2}$$

$$\beth(S_i) = \left(\mathfrak{k}_{i1}, \ldots, \mathfrak{k}_{ij}, \ldots, \mathfrak{k}_{i\ell}\right).$$

Each part of the information vector $\beth(S_i)$ is defined as

$$\mathfrak{k}_{ij} = \begin{cases} 1, \text{if} S_i \in \mathcal{C}_j; \\ 0, \text{if } S_i \notin \mathcal{C}_j. \end{cases}$$

3 Statement of the Problem

Let's look at the recognition object (RO) problem in its usual setup [1, 30, 37]. Assume we're given a collection of objects $\tilde{S}^q(\tilde{S}^q = \{S'_1, \ldots, S_{i'}, \ldots, S_{q'}\}, \tilde{S}^q \subset \mathbb{X})$ which are described in the feature space$\mathcal{X}$. The objective is to create RA, or model$\mathcal{A}$, that use sthestart inginformation, $\mathcal{I}_0$ to work out the values of the predicate $\mathrm{P_j}(S'_\mathrm{i})$ $\left(\mathrm{P_j}(S'_\mathrm{i}) = "S'_\mathrm{i} \in \mathcal{C}_\mathrm{j}"\right)$ for every object in$\tilde{S}^\mathrm{q}$:

$\mathcal{A}(\mathcal{I}_0, \tilde{S}^\mathrm{q}) = \|\nabla_{\mathrm{ij}}\|_{\mathrm{m}\times\|}, \nabla_{\mathrm{ij}} = \mathrm{P_j}(S'_\mathrm{i}),$

$$\mathrm{P_j}(S'_\mathrm{i}) = \begin{cases} 1, \text{ if the object } S'_\mathrm{i} \text{ belongs to the class } \mathcal{C}_\mathrm{j}; \\ 0, \text{ if the object } S'_\mathrm{i} \text{ does not belong to class } \mathcal{C}_\mathrm{j}; \\ \Delta, \text{ if the model did not calculate the values } \mathrm{P_j}(S'_\mathrm{i}). \end{cases}$$

4 Method of Solution

This study suggests a fresh method for resolving the task of building RAs specifically for categorizing objects defined in the HDFS. Following this method, an RAs has been created, relying on the building of OTRs. The primary concept behind this group of RAs involves forming Transform-Pattern Functions (TPFs) and later computing object similarity assessments. The key stages involved in the proposed RA family are outlined as follows.

1. **Formation of groups with strongly linked features (GSRF).** Here, a number $\mathrm{n}'(\mathrm{n}' < \mathrm{n})$ of "separate" GSRFs is produced. Thus, in the beginning, the parameter n' is determined, and its value is set when solving particular problems [35–38].
2. **Choosing a set of RFs.** This stage results in selecting one representative from each SRF, embodying as a typical element of its feature group. As a result, n' RFs are determined, which are described by an n-dimensional Boolean vector (BV) $\mathbb{r}$:

$$\mathbb{r} = (\mathrm{r}_1, .., \mathrm{r}_\mathrm{i}, .., \mathrm{r}_\mathrm{n}), \mathrm{n}' = \|\mathbb{r}\|_\mathrm{B}$$

where $\mathrm{r_i} = 1$, if the feature x_i is a representative, or $\mathrm{r_i} = 0$ otherwise.

Thus, consider the produced space of RFs as $\mathbb{X}'$ (where the dimension of $\mathbb{X}'$ equals n'). Therefore, choosing a group of RFs involves setting the vector $\mathbb{r}$elements, with the vector considered a parameter here [36–38].
3. **Calculating the DF in the SRF.** Well established dissimilarity metric, designed to measure how distinct objects $\mathrm{S_u}$ and S are from each other, all with in the spatial reference frame (SRF) $\mathbb{X}'$. Let's focus on two objects, $\mathrm{S_u}$ and S, positioned in the space $\mathbb{X}'$:

$\mathrm{S_u} = (\mathrm{a_{ui_1}}, \ldots, \mathrm{a_{ui_q}}, \ldots, \mathrm{a_{ui_{n'}}})$ and $\mathrm{S} = (\mathrm{a_{i_1}}, \ldots, \mathrm{a_{i_q}}, \ldots, \mathrm{a_{i_{n'}}})$.

The variance exhibited between these objects is formulated as:

$$\mathrm{d}(\mathrm{S_u}, \mathrm{S}) = \sum_{\mathrm{q}=1}^{\mathrm{n}'} \lambda_\mathrm{q}\rho\left(\mathrm{a_{ui_q}}, \mathrm{a_{i_q}}\right), \tag{3}$$

where $\lambda_1, \ldots, \lambda_q, \ldots, \lambda_{n'}$ are the RA parameters; $\rho\left(a_{ui_q}, a_{i_q}\right)$ – assessment of the difference between objects S_u and S calculated using the attribute x'_{i_q}. Thus, at this stage, set $\lambda_q\left(q = \overline{1, n'}\right)$ as a parameter, which determines the DF, presented in the form (3).

4. **Identification of SIOs in the SRF.** At this stage, sets $\mathbb{V}_A$ consisting of m' elements are determined. Moreover, each element $\mathbb{V}_A$ is a separate "independent" SIOs. The clustering process involves comparing objects from the initial data in pairs to assess their proximity inside the SRF. This results in m' distinct clusters:

$$\mathbb{V}_A = \{\mathfrak{V}_1, \ldots, \mathfrak{V}_q, \ldots, \mathfrak{V}_{m'}\}; \mathbb{V}_A \subset \mathbb{V}, \tag{4}$$

Observe that the formation of (4) employs function (3). Hence, during this stage, m' is predetermined, its value derived from an analysis of the training data.

5. **Defining BOs in the SRF.** Here, a set of BOs, E_q is defined that serve as representative examples from each strongly related pseudo-object with in the RF's k-dimensional subspace. The BO E_q components in each cluster of strongly connected objects are calculated by averaging all members of the $\mathfrak{V}_q$ cluster $\left(q = \overline{1, m'}\right)$:

$$b_{qi} = \frac{1}{|\mathfrak{V}_q|} \sum_{a_{ui} \in \mathfrak{V}_q} a_{ui},\ i = \overline{1, n'}. \tag{5}$$

Upon completion of this step, we possess m' BOs:
$\mathbb{E} = \{E_1, \ldots, E_q, \ldots, E_{m'}\}$.
Define Σ as the set encompassing all conceivable object groupings (subsets):

$$\Sigma = \{\sigma_{i_1}, \ldots, \sigma_{i_q}, \ldots, \sigma_{i_{m'}} | \sigma_{i_1} \in \Lambda_1, \ldots, \sigma_{i_q} \in \Lambda_q, \ldots, \sigma_{i_{m'}} \in \Lambda_{i_{m'}}\}.$$

Every element with in Σ corresponds to a single BV $\tilde{\sigma}$.The collection of BVs associated with all elements of Σ is denoted as $\{\tilde{\sigma}\}$. Consequently, the representative objects are characterized by the BV $\tilde{\sigma}$ ($\tilde{\sigma} \in \{\tilde{\sigma}\}$). Therefore, choosing a specific collection of base objects equates to finding the vector $\tilde{\sigma}$.With in this particular step, the vector $\tilde{\sigma}$ is treated as a pre-defined parameter.

6. **Construction of TPFs.** At this point, the proximity function (PF) is formulated for the base object S_u and the object S. With a group of selected BOs $\tilde{S}^{m'}$:

$$\tilde{S}^{m'} = \{S'_1, \ldots, S'_n, \ldots, S'_m\},$$

and the objects of the training set $\tilde{S}^m$ are calculated S_u using the i-th feature and a numerical sequence is obtained that characterizes the change in the assessment of the proximity of the base object S_u and the objects of the training set:

$$\mathcal{P}_i(S'_u, S_1), \mathcal{P}_i(S'_u, S_2), \ldots, \mathcal{P}_i(S'_u, S_v), \ldots, \mathcal{P}_i(S'_u, S_m), \tag{6}$$

where $\mathcal{P}_i(S'_u, S_v) = |a'_{iu} - a_{iv}|$.

Let us arrange all the numbers (6) that reflect the change in proximity assessment in order of its increase: from smallest to largest:

$$\mathcal{P}_i(S'_u, S_{v1}) \leq \mathcal{P}_i(S'_u, S_{v2}), \ldots, \mathcal{P}_i(S'_u, S_{vm}). \tag{7}$$

Each value of a numerical proximity score in a series represents a variant. If you arrange all the values of the numerical estimates in ascending order, you will get a variation series.

Based on the analysis of (7), we determine ℓ' the subsets $\mathfrak{D}_{i1} \ldots \mathfrak{D}_{ik} \ldots \mathfrak{D}_{i\ell'}$. In each subset such objects accumulate, the distances between these objects and the base object S_u is located in a certain interval. The elements of each subset, according to the th attribute under consideration, are located close to the elements of other subsets, i.e., from $\mathcal{P}_i(S'_u, S_v) < \mathcal{P}_i(S'_u, S_u)$(under the condition that the object S_v belongs to the subset $\mathfrak{D}_k$ and the object S_u does not) it follows that $P_{uk}(a_{iv}) \bigwedge P_{uk}(a_{iu}) = 0$. Here is $P_{uk}(a_{iv})-$ a predicate characterizing the belonging of the u th object to the subset $\mathfrak{D}_{ik}$ according to the i th attribute:

$$P_{uk}(a_{iv}) = \begin{cases} 1, \textit{if } a_{iv} \in \mathfrak{D}_{ik}; \\ 0, \textit{if } a_{iv} \notin \mathfrak{D}_{ik}. \end{cases}$$

Next, the number of occurrences of objects of each class in the considered subset (group) of objects obtained by assessing the proximity between the base object S_u and the objects of the training sample is determined. In this case, proximity scores are calculated for each attribute. Based on the appraisal of the proximity, calculated by attribute, between the base object S_u and the objects of the training sample, we will construct Table 1.

Table 1. Distribution of proximity estimates between the base object S_u and objects of the training sample, calculated using the i th feature.

Distance Class	Group 1	Group k	Group ℓ'
C_1	m_{iu11}	m_{iu1k}	$m_{iu1\ell'}$
…	…	…	…
C_j	m_{iuj1}	m_{iujk}	$m_{iuj\ell'}$
…	…	…	…
C_ℓ	$m_{iu\ell 1}$	$m_{iu\ell k}$	$m_{iu\ell\ell'}$
Sum	μ_{iu1}	μ_{iuk}	$\mu_{iu\ell'}$

After determining the number of occurrences of objects, you can determine the frequency of existence of objects of each class in the considered subset (groups) of objects. Let μ_{iuk} be the number of objects belonging $k-$ to the subset $\mathfrak{D}_{ik}$, and $\underset{\sim}{m_{iujk}}$ let be the number of objects belonging to the subsets $\mathfrak{D}_{ik}$ and C_j. Then the frequency of occurrence of objects of each class in the considered subset (groups) of objects relative to the base object S_u is determined as follows:

$$c_{iujk} = \frac{m_{iujk}}{\mu_{iuk}}.$$

Based on the assessment of the frequency of existence of objects of each class in the considered subset (groups) of objects relative to the base object. Calculated by the *i*-th attribute, between the base object S_u and the objects of the training sample, we will construct TPFs between the base object S_u and the object S(see Table 2).

Table 2. Tabular PF, constructed as an estimate of the frequency of occurrence of objects of each class in the considered subset (groups) of objects relative to the base object, calculated using the *i* th characteristic

Distance	Group 1	Group k	Group ℓ'
C_1	c_{iu11}	c_{iu1k}	$c_{iu1\ell'}$
...	...	...	...
C_j	c_{iuj1}	c_{iujk}	$c_{iuj\ell'}$
...	...	...	...
C_ℓ	$c_{iu\ell 1}$	$c_{iu\ell k}$	$c_{iu\ell\ell'}$

Thus, as a result of execution, a tabular PF is formed for each feature. This function is characterized n', m', ℓ, ℓ' by parameters. Typically, the parameters n', m', ℓ' receive small values.

7. **Calculating the PSB the base object S_u and the object S in all respects.** At this stage of finding the RA family a numerical characteristic called a score is specified. The estimate is determined by the values of the tabulated PF given in Table 2. Note that the table function characterizes the degree of proximity between the base object S_u and the object S within the *j* th attribute. In addition, the score may depend on the weight parameters of each feature calculated for each subset $\mathcal{D}_{iuk}$. Then the PSB the base object S_u and the object S in the space of RFs is calculated as follows:

$$\mathcal{E}_j(S_u, S) = \sum_{i=1}^{n'} \sum_{k=1}^{l'} \hbar_{ik} c_{iujk}, \tag{8}$$

where $\hbar_{ik}$–weight parameter i– of the th attribute.

Thus, at this stage, the DF (8) is specified, which quantifies the difference between objects. In this case, function (8) is determined by the parameters $n' \times l'$.

8. **Calculation of the PSB the class C_j and the object S.** The assessment of an object's class C_j membership S($j = \overline{1, l}$) is calculated as follows:

$$\mathfrak{B}(S) = (\mathfrak{G}_1(S), \ldots, \mathfrak{E}_j(S), \ldots, \mathfrak{G}_l(S)),$$

$$\mathfrak{E}_j(S) = \sum_{S_u \in \tilde{S}^{m'}} \gamma_u \mathcal{E}_j(S_u, S),$$

where γ_u is the RA parameter ($u = 1, \ldots, m'$).

Therefore, we've established a group of RA built upon one-dimensional threshold rules. Every algorithm $\mathfrak{B}$ from this model is fully characterized by parameters $\widetilde{\pi}$. The collection of all RA with in this group is designated as $\mathfrak{B}(\widetilde{\pi}, S)$. Finding the most effective recognizing operator within this group is done within the parameter space $\widetilde{\pi}$. The optimal operator $\mathfrak{B}(\widetilde{\pi}_0, S)$ is chosen by looking for the minimum value of the quality functional of the RAs:

$$\mathfrak{R}\left(\widetilde{\pi}, \tilde{S}^m\right) = \Theta\left(\mathfrak{B}\left(\widetilde{\pi}, \tilde{S}^m\right)\right)/\mathfrak{m},$$

$$\Theta\left(\mathfrak{B}\left(\widetilde{\pi}, \tilde{S}^m\right)\right) = \left(\sum_{S\in\tilde{S}^m} \mathfrak{Q}\left(\| \widetilde{\alpha}\ (S) - \mathfrak{C}\left(\mathfrak{B}(\widetilde{\pi}, S)\right)\|\right)\right),$$

$$\mathfrak{m} = \left|\tilde{S}^m\right|, \mathfrak{Q}(\S) = \begin{cases} 1, \mathit{if}\ x = 0; \\ 0, \mathit{if}\ x \neq 0, \end{cases}$$

where $\| \cdot \|$ is the norm of the BV.

To test the performance of the considered family of RA, it is necessary to provide experimental studies. The following is a description of the experimental studies carried out.

5 Experiments and Results

Functional diagrams of the proposed software complex and accompanying procedures have been built in order to carry out experimental tests to evaluate the examined family of RA. These processes are implemented as software in C++ using the OpenCV library.

While addressing a model problem, the performance of the suggested family of RAs was experimentally investigated. The RA models selected for testing are: 1) the conventional RA model using potential functions (A_1-family) [16]. 2) the proposed model (A_2-family). The fact that A_1 and A_2 are members of the same class of RA models explains why the are members of the same class of RA models explains why the A_1-family was selected for comparison.

We assessed the given RA models, judging how well they did on the set tasks using three measures: the proportion of correctly identified objects in the test data; how long the algorithm took to train (in seconds); and the time the algorithm spent identifying objects in the test data (also in seconds).

Consider an initial dataset T with m objects, represented as $\{S_1, \ldots, S_{\mathrm{u}}, \ldots, S_{\mathrm{m}}\}$. To assess quality based on the accuracy of evaluated RA models, the set $\mathrm{T}(\mathrm{T} = \{S_1, \ldots, S_{\mathrm{u}}, \ldots, S_{\mathrm{m}}\})$ is split into 2 subsets, $\mathrm{V_t}$ and $\mathrm{V_c}$ ($\mathrm{T} = \mathrm{V_t} \cup \mathrm{V_c}$, $\mathrm{V_t}$ being the training subset, V_c the control subset).To prevent biased (either good or bad) splits of T into 2 parts, we use the sliding control method [6, 38, 39]. Essentially, the original item set T is randomly divided subsets. This results in the sets $\mathbb{T}$:

$$\mathbb{T} = \left\{\mathcal{T}_1, \ldots, \mathcal{T}_{\mathrm{u}}, \ldots, \mathcal{T}_f\right\}, |\mathcal{T}_{\mathrm{u}}| = \sum_{\mathrm{j}=1}^{1}\left|\mathcal{T}_{\mathrm{u}} \cap \mathcal{C}_{\mathrm{j}}\right|.$$

In this case, it is required that the elements of the set $\mathbb{T}$ satisfy the following simple conditions (for u, v $\in \{1, \ldots \int\}, \int = 10$):

1) $\mathcal{T}_u \cap \mathcal{T}_v = \varnothing, u \neq v$;
2) $|\mathcal{T}_1| = \cdots = |\mathcal{T}_u| = \cdots = |\mathcal{T}_\int|$;
3) $|\mathcal{T}_1 \cap \mathcal{C}_j| \approx \cdots \approx |\mathcal{T}_u \cap \mathcal{C}_j| \approx \cdots \approx |\mathcal{T}_\int \cap \mathcal{C}_j|$.

The key attributes of the original data created for the model problem in this test are: an initial sample size of $m = 1000$, two classes ($\ell = 2$), 500 features ($n = 500$), and six GSRFs ($n' = 6$). The sizes of the training and control maples were $|V_t| = 900$, $|V_c| = 100$.

As stated previously, the model task was addressed with the help of the RA A_1 and A_2. The accuracy during the training phase was 95.2% for A_1 and 94.3% for A_2.When using these RAs in the testing phase, the outcomes for solving the specified task were 78.9% and 89.1%, respectively (see Table 3).

Table 3. Results of solving the model problem

RA family	Time, s		Recognition accuracy, %
	Education	Recognition	
$\mathcal{A}_1$-family	6, 253	0.011	78.9
$\mathcal{A}_2$-family	10,674	0.003	89.1

By analyzing the results, it's clear that the proposed RA model $\mathcal{A}_2$ improved the identification accuracy of items in the feature space by more than 10% compared to $\mathcal{A}_1$. This improvement stems from $\mathcal{A}_2$ utilization of several methods designed to boost recognition accuracy, including feature selection, determining DF in the representative feature subspace, and creating a tabular PF. The experimental results suggest that when the training data volume and the size of the feature space are considerable, the suggested RA model addresses the RO issue more effectively.

6 Conclusions

During the study, a novel technique was introduced, using OTR creation to categorize elements within the HDFS. With this technique, transitioning from the initial HDFS to a reduced-dimension RF environment is possible. This has resulted in the development of an RA model based on the OTR framework. The central principle of the proposed model involves building TPFs from a selection of fundamental items.

The experimental results revealed that the proposed RA model improves recognition accuracy and significantly decreases the computational load for identifying an unknown object in the HDFS.

Creating a powerful algorithm also caused training time to go up. This can be understood because, unlike training a standard RA model, this one uses complex optimization methods during training.

When we worked on a variety of test problems, especially those in the Experiments and Results section, it became clear that choosing BOs, especially determining their size and building TPFs, are critical for building a robust RA. Therefore, future research should focus on finding the best BOs and building TPFs.

References

1. Zhuravlev, Yu.I.: An algebraic approach to recognition or classifications problems. Pattern Recognit Image Anal. **8**(1), 59–100 (1998)
2. Homenda, W., Pedrycz, W.: Pattern Recognition: A Quality of Data Perspective. Wiley, New York (2018)
3. Beyere, M., Richter, M., Nagel, M.: Pattern Recognition: Introduction, Features, Classifiers and Principles. De Gruyter Oldenbourg, Boston (2018)
4. Zhuravlev, Yu. I.: Selected Scientic Works. Magister, Moscow (1998)
5. Fazilov, Sh., Radjabov, S., Mirzaev, O., Mirzaeva, S.: Construction of the model of recognition operators in the large dimensional feature space. J. Phys. Conf. Ser. **1210**, 012044 (2019). https://doi.org/10.1088/1742-6596/1210/1/012044
6. Rasulmukhamedov, M., Mirzaev, O., Khakimov, Sh., Shukurov, F.: Model of recognition algorithms based on the selection of basic objects. AIP Conf. Proc. **2789**, 040099 (2023). https://doi.org/10.1063/5.0145607. PTLICISIWS-2022
7. McLachlan, G.J.: Discriminant Analysis and Statistical Pattern Recognition. Wiley, New York (2004)
8. Zhuravlev, Yu. I., Dyusembaev, A.E.: Neural network construction for recognition problems with standard information on the basis of a model of algorithms with piecewise linear surfaces and parameters. Dokl. Math. **100** (2), 411–415 (2019)
9. Tou, J., Gonzalez, R.: Pattern Recognition Principles. Mir, Moscow (1978)
10. Li, Y., Liu, B., Yu, Y., Li, H., Sun, J., Cui, J.: 3E-LDA: three enhancements to linear discriminant analysis. In: ACM Trans. on Knowledge Discovery from Data, pp. 1–20. https://doi.org/10.1145/3442347
11. Li, C., Shao, Y., Yin, W., Liu, M.: Robust and sparse linear discriminant analysis via an alternating direction method of multipliers. IEEE Trans. Neural Netw. Learn. Syst. **31** (3), 915–926 (2020). https://doi.org/10.1109/TNNLS.2019.2910991
12. Duda, R., Hart, P., Stork, D.: Pattern Classification. Wiley, New York (2001)
13. Webb, R.A., Copsey, K.D.: Statistical Pattern Recognition. Wiley, New York (2011)
14. Jain, A.K., Duin, R.P.W., Mao, J.: Statistical pattern recognition: a review. IEEE Trans. Pattern Anal. Mach. Intell. **22** (1), 4–37 (2000). https://doi.org/10.1109/34.824819
15. Merkov, A.B.: Pattern Recognition: An Introduction to Statistical Learning Methods. URSS, Moscow (2019)
16. Ayzerman, M.A., Braverman, E.M., Rozonoer, L.I.: Method of Potential Functions in the Theory of Machine Learning. Nauka, Moscow (1970)
17. Dubrovin, V.I., Koretsky, N.Kh., Subbotin, S.A.: Modified method of potential functions. Complex Syst. Process. **1**, 12–19 (2002)
18. Oliveri, P.: Potential function methods: efficient probabilistic approaches to model complex data distributions. NIR News SAGE J. **28**(4), 14–15 (2017). https://doi.org/10.1177/0960336017703253
19. Pavlov, Y.: Potential function method and stochastic approximation, preferences and value evaluation. https://doi.org/10.13140/RG.2.2.21234.76489

20. Sulewski, P.: Potential function method approach to pattern recognition applications. In: Environment. Technology. Resources, Proceedings of the 11th International Scientific and Practical Conference, Rezekne, Latvia, pp. 30–35 (2017). https://doi.org/10.17770/10.17770/etr2017vol2.2512
21. Sulewski, P.: Recognizing distributions using method of potential functions. Commun. Stat. Simul. Comput. (2021). https://doi.org/10.1080/03610918.2021.1908561
22. Kudryavtsev, V.B., Andreev, A.E., Hasanov, E.E.: Test Recognition Theory. Fizmatlit, Moscow (2007)
23. Lbov, G.S., Startseva, N.G.: Logical Decision Functions and Questions of Statistical Stability of Decisions. IM SB RAS, Novosibirsk (1999)
24. Djukova, E.V., Masliakov, G.O., Prokofyev, P.A.: On the logical analysis of data with partial orders in the problem of classification by precedents. J. Comput. Math. Math. Phys. **59**(9), 1605–1616 (2019)
25. Djukova, E.V., Masliakov, G.O., Prokofyev, P.A.: Logical classification of partially ordered data. In: Kuznetsov, S., Panov, A. (eds.) Artificial Intelligence. RCAI 2019. Communications in Computer and Information Science, vol. 1093, pp. 115–126. Springer, Cham (2019). https://doi.org/10.1007/978-3-030-30763-9_10
26. Povkhan, I.: Logical recognition tree construction on the basis of a step-to-step elementary attribute selection. Radio Electron. Comput. Sci. Control **2**, 95–105 (2020). https://doi.org/10.15588/1607-3274-2020-2-10
27. Povkhan, I.: Logical classification trees in recognition problems. In: Informatyka, Automatyka, Pomiary w Gospodarce i Ochronie Środowiska (IAPGOS), vol. 10, no. 2, pp. 12–15 (2020). https://doi.org/10.35784/iapgos.927
28. Ignat'ev, O.A.: Construction of a correct combination of estimation algorithms adjusted using the cross validation technique. Comput. Math. Math. Phys. **55**(12), 2094–2099 (2015). https://doi.org/10.1134/S0965542515120064
29. Nishanov, A.K., Djurayev, G.P., Khasanova, M.A.: Improved algorithms for calculating evaluations in processing medical data. COMPUSOFT International Journal of Advanced Computer Technology **8** (6), 3158–3165 (2019)
30. Zhuravlev, Yu.I., Ryazanov, V.V., Senko, O.V.: Recognition. Mathematical Methods. Software System. Practical Applications. Fazis, Moscow (2006)
31. Kabulov, A., Babadzhanov, A., Saymanov, I.: Correct models of families of algorithms for calculating estimates. AIP Conf. Proc. Appl. Math. Inf. Technol. **2781**(1), 020010 (2023). https://doi.org/10.1063/5.0144830\
32. D'yakonov, A.G.: Theory of equivalence systems for describing algebraic closures of a generalized estimation model. Comput. Math. Math. Phys. **50**(2), 369–381 (2010). https://doi.org/10.1134/S0965542510020181
33. D'yakonov, A.G.: Theory of equivalence systems for describing algebraic closures of a generalized estimation model. II, Comput. Math. Math. Phys. **51**(3), 490–504 (2011). https://doi.org/10.1134/S0965542511030067
34. Lantz, B.: Machine Learning with R: Expert Techniques for Predictive Modeling. Packt Publishing Ltd (2019)
35. Mirzaev, O., Radjabov, S., Mirzaev, N., Rabbimov, I., Baratov, J.: Construction of statistical recognition algorithms based on two-dimensional threshold functions. Procedia Comput. Sci. **234**, 123–130 (2024). https://doi.org/10.1016/j.procs.2024.02.158
36. Mirzaev, O., Radjabov, S., Mirzaev, N., Meliev, F., Tillavoldiev, A.: A family of recognition algorithms based on the construction of k-dimensional threshold rules. Procedia Comput. Sci. **237**, 610–617 (2024). https://doi.org/10.1016/j.procs.2024.05.146
37. Fazilov, Sh., Mirzaev, N., Mirzaeva, G.: Modified recognition algorithms based on the construction of models of elementary transformations. Procedia Comput. Sci. **150**, 671–678 (2019). https://doi.org/10.1016/j.procs.2019.02.037

38. Fazilov, Sh., Mirzaev, N., Radjabov, S., Mirzaev, O.: Determining of parameters in the construction of recognition operators in conditions of features correlations. In: Proceedings of the School-Seminar on Optimization Problems and their Applications (OPTA-SCL 2018), vol. 2098 (2018). http://ceur-ws.org/Vol-2098/paper10.pdf
39. Braga-Neto, U.M., Dougherty, E.R.: Error Estimation for Pattern Recognition. Springer, New York (2016)

Search of Neural Network Architecture for the Problem of Estimating Lipophilicity of Small Organic Compounds

Boris Piakillia(✉) and Valerii Goncharov

Tomsk Polytechnic University, Tomsk, Russia
morphism@tpu.ru

Abstract. In modern chemistry, lipophilicity, which refers to the affinity of a compound for lipids or fats, is recognized as one of the crucial parameters influencing the behavior and properties of organic compounds. This parameter is particularly important because it affects a compound's solubility, permeability, and distribution within biological systems, thereby playing a vital role in drug design and development. Traditional methods for estimating lipophilicity typically involve experimental techniques such as partition coefficient measurements, which are often time-consuming and resource-intensive. These methods require extensive laboratory work, the use of sophisticated equipment, and a significant amount of reagents, leading to high costs and prolonged timelines. Additionally, the accuracy of these experimental methods can be influenced by various factors, including the purity of the compounds and the precision of the experimental conditions. In response to these limitations, the field has increasingly turned to computational approaches, particularly machine learning techniques, which offer a more efficient and cost-effective alternative for predicting lipophilicity. The primary objective of this study is to enhance the prediction accuracy of lipophilicity estimates by optimizing the architecture of neural networks, a subset of machine learning models. In this research, we focus on systematically optimizing these architectural parameters to develop a neural network model that provides reliable and precise lipophilicity predictions for a wide range of organic compounds.

Ultimately, the goal is to create a robust computational tool that can significantly reduce the time and resources required for lipophilicity estimation, facilitating faster and more efficient drug development processes. By leveraging the power of machine learning, we aim to advance the capabilities of computational chemistry and contribute to the development of novel organic compounds with desirable properties.

Keywords: Modeling · Neural Network · Lipophilicity · Cheminformatics · Machine Learning · Artificial Intelligence · Hyperparameter Optimization

1 Introduction

In the fields of chemistry and pharmacology, it is now widely recognized that the lipophilicity of organic compounds is a critical factor influencing their properties and biological activities [1]. Lipophilicity, which describes the affinity of a compound for

V. Jordan et al. (Eds.): HPCST 2024, CCIS 2919, pp. 128–136, 2026.
https://doi.org/10.1007/978-3-032-20325-0_10

lipid environments over aqueous ones, plays a significant role in determining the absorption, distribution, metabolism, excretion, and toxicity (ADMET) of drugs. Despite its importance, traditional methods for measuring lipophilicity, both experimental and theoretical, often face substantial limitations. These methods can be time-consuming and resource-intensive, and they frequently struggle to accommodate novel or structurally complex compounds, limiting their applicability in rapidly advancing fields [2].

In this context, the advent of machine learning presents a promising alternative for predicting lipophilicity from the structural data of molecules. The primary goal of this study is to develop and refine a neural network architecture to enhance the accuracy and efficiency of lipophilicity estimation for organic compounds. The study aims to leverage the power of machine learning to overcome the limitations of traditional methods, providing a faster and more flexible approach to lipophilicity prediction.

Significant research has been conducted on the application of neural networks and machine learning in chemistry and pharmacology, particularly in estimating lipophilicity. Deep learning methods, including graph-based neural networks, have been at the forefront of this research. For instance, a study published in the Journal of Cheminformatics introduced a neural network model designed to predict lipophilicity and water solubility. In this model, each chemical compound is represented as a mathematical graph, capturing the intricate relationships between atoms and bonds within the molecule [3]. Moreover, other studies have explored the use of advanced deep neural network models, such as the "transformer" architecture, to predict various molecular properties [4]. These studies underscore the potential of machine learning to address complex problems in chemistry and pharmacology. They highlight the critical role of selecting and adapting the model architecture to achieve high accuracy and efficiency in predictions. However, a significant challenge remains: processing and representing chemical structures as graphs require considerable computational resources and specialized knowledge in cheminformatics.

Additionally, the performance of these models is highly dependent on the quality and completeness of the input data. In the realm of pharmacology, the availability of comprehensive chemical data is often limited, posing a major obstacle to the effectiveness of machine learning models. Accurate predictions rely on high-quality data, and the scarcity of such data can hinder the development and validation of robust models.

In summary, while traditional methods for measuring lipophilicity have significant limitations, machine learning offers a promising alternative that could revolutionize this field. By developing and adapting neural network architectures, this study aims to improve the accuracy and efficiency of lipophilicity predictions. Despite the challenges of computational demands and data quality, the advancements in machine learning techniques hold great potential for enhancing our understanding and prediction of molecular properties in chemistry and pharmacology.

2 Problem Statement and Source Data

The present article focuses on the task of finding the architecture of a fully connected neural network for assessing lipophilicity, which is a key element in the development of new pharmaceutical drugs and chemical compounds. During the training of neural networks, an open dataset on lipophilicity was used, taken from the ChEMBL database [5], which contains information on chemical compounds and their biological activity. Lipophilicity is measured by a dimensionless logarithmic index that shows a molecule's ability to associate with fats and ranges from -2 (high hydrophilicity) to 5 (high lipophilicity) [6]. This dataset includes the following parameters:

- • 4200 organic compounds.
- • The average lipophilicity is 2.18.
- • The standard deviation of lipophilicity is 1.20.

The data will be split into training and testing sets in a 90 to 10 ratio. The training set is intended for training and cross-validation of the neural network, while the test set will be used to evaluate its performance on previously unseen data. This approach to data splitting is widely accepted and recommended for assessing the model's ability to effectively work with new data [7]. To ensure the reproducibility of results, the initial value of the random number generator (random seed) is set to 42. A random seed is an initial number that serves as a starting point for generating a sequence of random numbers. Assigning a specific value to the random seed guarantees the reproducibility of processes such as data splitting, initialization of weights in the model, and other operations using the random number generator with each rerun of the model. The Fig. 1 shows.

For the task of predicting lipophilicity LogP, which is a regression task, the optimal choice of metric is usually the Mean Squared Error (MSE) or the Root Mean Squared Error (RMSE) [7]:

$$\mathrm{RMSE} = \sqrt{\frac{1}{N} \cdot \sum_{i=1}^{N} (y - \hat{y})^2} \tag{1}$$

where y represents the experimental values of lipophilicity LogP, and $\hat{y}$ represents the values of lipophilicity predicted by the model.

In this work, we will take the root mean square error (1) as the main metric due to its better interpretability compared to MSE. This choice is explained by the fact that the RMSE error is expressed in the original units of measurement of lipophilicity LogP, which is a logarithmic quantity. In cheminformatics, various chemical descriptors and fingerprints are often used to describe molecules, which facilitate the analysis and comparison of molecular structures, as well as the prediction of their properties and biological activity. The most common features include:

- Molecular descriptors such as molecular weight, number of hydrogen donors and acceptors, molecular surface area, moment of inertia, and others. These descriptors are valuable tools due to their simplicity and broad applicability. They are easy to compute, provide intuitive insights, and correlate well with properties like solubility and bioavailability, enhancing predictive accuracy in machine learning models. Their computational efficiency further adds to their appeal. However, these descriptors can oversimplify complex molecular properties, potentially missing critical information and dynamic behaviors. Their relevance varies with context, and they depend heavily on the quality of input data. High correlation among descriptors can lead to redundancy, complicating analysis and model development;

- MACCS keys, a standard set of 166 bits that reflect the presence or absence of certain chemical structures or patterns in a molecule. These predefined structural keys provide a standardized way of representing molecular structures, making them simple to use and understand. They are widely accepted and used in the industry, offering a common language for comparing and sharing molecular data across different platforms and studies. Their computational efficiency is another significant advantage, as generating MACCS keys is relatively fast and requires minimal computational resources. Furthermore, they are effective for certain types of similarity searches and structure-activity relationship (SAR) studies, providing useful insights into the presence or absence of specific functional groups and substructures. However, MACCS keys also come with limitations. One major drawback is their binary nature, which can oversimplify complex molecular structures and interactions. This binary representation may lead to a loss of nuanced information, reducing the accuracy of predictions in some cases. Additionally, MACCS keys have a fixed size and predefined set of structural features, which might not capture all relevant aspects of diverse chemical spaces, limiting their applicability for novel or unconventional compounds. Their reliance on the presence or absence of specific substructures can also result in high sensitivity to minor structural variations, sometimes leading to misleading similarity assessments;
- Extended Connectivity Fingerprints (ECFP), structure-based fingerprints that consider the environment of each atom. These fingerprints are highly informative and versatile. They capture detailed information about molecular structures by considering the connectivity of atoms, making them effective for a wide range of applications, including similarity searches, virtual screening, and quantitative structure-activity relationship (QSAR) modeling. ECFPs are also flexible, as their design allows for varying the radius to capture different levels of structural detail. This adaptability enables the representation of both local and more extended structural features. Additionally, ECFPs are robust against minor variations in molecular structures, which enhances their reliability in similarity assessments. Their ability to capture complex structural information makes them particularly valuable for identifying bioisosteres and scaffold hopping in drug discovery;
- Continuous Distributed Description of Drug-like molecules (CDDD) is a method that uses deep learning to transform molecular structures into a continuous vector space [8]. Offers significant advantages in cheminformatics due to its continuous and distributed representation of molecular structures, which enhances accuracy and generalization in predictive models. By leveraging deep learning, CDDD can automatically learn relevant features from molecular data, reducing bias and handling large, diverse chemical spaces effectively. This makes it highly suitable for virtual screening, QSAR modeling, and drug discovery. However, CDDD has several limitations. It requires substantial computational resources and expertise in machine learning, which can be a barrier for some researchers. The high-dimensional embeddings produced by deep learning models are less interpretable than traditional descriptors, complicating the

understanding of molecular features and biological activity relationships. The performance of CDDD models is also highly dependent on the quality and quantity of training data, necessitating careful tuning and rigorous evaluation.

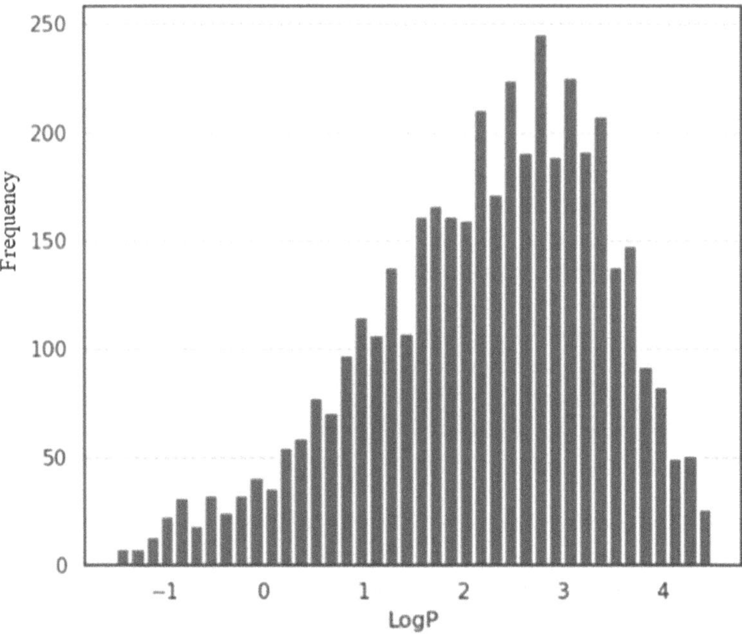

Fig. 1. Graph of lipophilicity distribution in ChEMBL data

In this study, ECFP chemical fingerprints, which are a standard in most cheminformatics studies [7], were chosen as input features to describe chemical compounds. They capture detailed structural information by considering the connectivity of atoms, which allows for a nuanced understanding of molecular properties and interactions. This flexibility, including the ability to adjust the radius, makes ECFPs adaptable to various levels of detail required in different applications. ECFPs are robust to minor structural variations, ensuring reliability in similarity assessments and facilitating tasks like bioisostere identification and scaffold hopping in drug discovery. Their proven utility in QSAR modeling and virtual screening demonstrates their effectiveness in predicting molecular activity, often correlating well with biological properties. ECFPs strike a balance between computational efficiency and informational richness, making them feasible for large-scale studies without the extensive resources needed for deep learning methods like CDDD.

Furthermore, ECFPs are widely accepted in the cheminformatics community, facilitating comparison and validation across different studies. Their versatility and effectiveness in tasks such as similarity searches, virtual screening, and feature selection in machine learning models make ECFPs a strong choice for molecular feature representation in cheminformatics.

3 Learning Process

The learning process of a neural network entails not only selecting an appropriate feature space but also establishing conditions that enable the model to perform optimally on the test set while avoiding overfitting. For this purpose, a fully connected neural network will serve as the foundational model. The input dimension of this network is dictated by the chosen set of features. The network architecture consists of two hidden layers: the first layer comprises 64 neurons, while the second layer contains 32 neurons. The output layer is designed with a single neuron, as the objective is to predict a single value—lipophilicity.

The activation function employed for the hidden layers is the Rectified Linear Unit (ReLU). ReLU is selected due to its ability to mitigate the vanishing gradient problem, which is a common issue in networks utilizing sigmoid or tangent activation functions. This ensures that the gradients remain substantial enough for efficient learning during backpropagation, thereby enhancing the overall training process.

Optimization of the neural network is achieved using the Adam (Adaptive Moment Estimation) method. Adam is particularly effective because it adapts the learning rate for each parameter individually, combining the advantages of two other extensions of stochastic gradient descent: AdaGrad and RMSProp [9, 10]. This results in faster convergence and improved performance on a variety of tasks.

The chosen architecture of this fully connected neural network is computationally efficient and designed to minimize the risk of overfitting. Overfitting is a critical concern, especially in the context of cheminformatics, where data collection can be expensive and labor-intensive. Unlike graph neural networks, which typically require a large number of parameters and extensive training data, this simpler fully connected network architecture helps ensure that the model remains manageable and cost-effective to train.

In summary, the architecture includes an input layer defined by the selected features, two hidden layers with 64 and 32 neurons respectively, and an output layer with a single neuron. The use of ReLU activation functions and the Adam optimizer enhances the training efficiency and overall model performance. This approach strikes a balance between computational simplicity and the necessity to avoid overfitting, making it particularly suitable for applications in cheminformatics where data resources are often limited [7, 9].

4 Search of Neural Network Hyperparameters

To determine the optimal architecture of the neural network for minimizing the test RMSE value, in conjunction with the application of selected Extended Connectivity Fingerprint (ECFP) features, the Grid Search method will be employed. This method involves an exhaustive search across a specified range of hyperparameters to identify the configuration that yields the best performance. Specifically, the Grid Search will systematically vary the number of neurons in each of the two hidden layers, exploring a range from 1 to 256 neurons per layer. This approach ensures a thorough exploration of possible architectures, facilitating the discovery of the most effective neural network configuration.

The choice of the neuron range from 1 to 256 is driven by the need to balance the model's complexity and the associated computational costs. As the number of neurons increases, the capacity of the network to learn intricate patterns within the data also increases. However, this comes at the cost of greater computational resources and the potential risk of overfitting, where the model learns the training data too well but performs poorly on unseen data. By exploring a broad range, we aim to identify a configuration that maximizes predictive performance while maintaining computational efficiency.

Grid Search operates by evaluating every possible combination of the specified hyperparameters within the given range. In this context, it will test all possible pairings of neuron counts for the two hidden layers. This exhaustive evaluation is crucial for ensuring that no potentially optimal configurations are overlooked. Each configuration is assessed based on its cross-validation RMSE value, a reliable indicator of the model's ability to generalize to new, unseen data.

The results of this Grid Search process are presented in Fig. 2. The figure illustrates the performance of different neural network architectures, highlighting the configuration that achieved the lowest cross-validation RMSE value. Notably, the optimal architecture identified consisted of 229 neurons in each of the two hidden layers. This configuration yielded a minimum cross-validation RMSE value of 0.786, indicating a strong predictive capability.

Further validation was performed by evaluating the test RMSE value for this optimal configuration. The test RMSE value was found to be 0.687, which is a substantial 10.3% improvement compared to the baseline network architecture. The baseline model, which included 64 neurons in the first hidden layer and 32 neurons in the second hidden layer, achieved a higher test RMSE value, demonstrating inferior predictive performance.

The significant reduction in RMSE achieved by the optimal configuration can be attributed to several factors. Firstly, the higher number of neurons allows the network to capture more complex and nuanced patterns within the ECFP features. ECFP features are known for their ability to represent intricate molecular structures and interactions, which are crucial for accurate predictions in cheminformatics applications. The larger network can better leverage these rich features to improve prediction accuracy.

Secondly, the exhaustive nature of the Grid Search ensures that the chosen configuration is indeed the best among the tested options. By systematically evaluating each possible combination, Grid Search minimizes the risk of settling on a suboptimal architecture. This thorough approach is particularly important in neural network design, where small changes in architecture can have significant impacts on performance.

Moreover, the improvement in performance underscores the importance of tailoring the neural network architecture to the specific characteristics of the data. While the baseline model provided a reasonable starting point, the optimal configuration identified through Grid Search demonstrated that a more customized approach could yield substantially better results. This highlights the need for detailed hyperparameter tuning in machine learning projects, especially when dealing with complex datasets and advanced feature representations like ECFP.

In conclusion, the use of Grid Search to determine the optimal neural network architecture for predicting molecular properties with ECFP features has proven highly effective. The method's exhaustive exploration of hyperparameter combinations allowed for

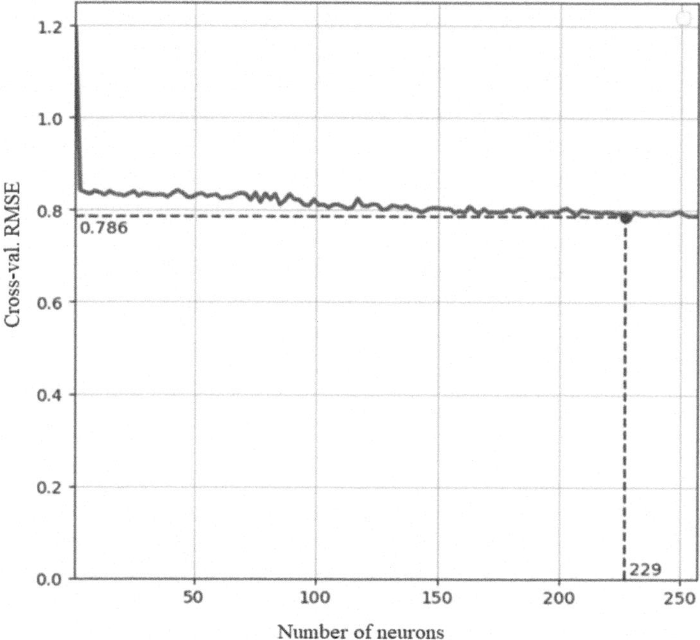

Fig. 2. Plot of the dependence of cross-validation RMSE on the number of neurons in the hidden layers.

the identification of a configuration that significantly outperformed the baseline model. This optimal architecture, with 229 neurons in each hidden layer, achieved lower RMSE values, demonstrating enhanced predictive accuracy. The success of this approach illustrates the critical role of careful hyperparameter tuning and the benefits of leveraging detailed molecular features in cheminformatics.

5 Conclusion

The study focused on optimizing the architecture of a fully connected neural network to predict lipophilicity using ECFP descriptors, underscoring the critical importance of selecting an appropriate architecture based on the given metric, which in this case was the root mean squared error (RMSE).

The adaptation of the network architecture involved meticulously determining the optimal number of neurons in the hidden layers, specifically considering the ECFP features. ECFP descriptors provide a detailed and nuanced representation of molecular structures by capturing the connectivity of atoms within a molecule. This detailed representation is crucial for accurately predicting complex properties like lipophilicity. By carefully tuning the number of neurons in the hidden layers, the model could more effectively learn and capture the intricate patterns and relationships within the ECFP features.

This architectural optimization led to a significant increase in prediction accuracy. The optimized model, tailored to leverage the detailed information provided by the

ECFP descriptors, outperformed the original, non-optimized model. The improvement in accuracy demonstrates how a well-chosen neural network architecture can better harness the richness of ECFP features, leading to more precise predictions.

This success highlights the broader principle that in cheminformatics, and machine learning in general, the architecture of a neural network must be carefully tailored to the specific characteristics of the input data. ECFP features, with their detailed molecular representations, require a neural network architecture that can adequately process and learn from this complexity. The study illustrates that optimizing the number of neurons and overall network structure allows the model to fully exploit the informative power of ECFP descriptors, thereby enhancing its predictive capabilities. This careful approach to model architecture selection is essential for improving the accuracy and reliability of predictive models in the field of cheminformatics.

References

1. Wardecki, D., Dołowy, M., Bober-Majnusz, K.: Assessment of lipophilicity parameters of antimicrobial and immunosuppressive compounds. Molecules **28**(6), 1–14 (2023)
2. Miller, R., Madeira, M., Wood, H., Geissler, W.: Integrating the impact of lipophilicity on potency and pharmacokinetic parameters enables the use of diverse chemical space during small molecule drug optimization. J. Med. Chem. **63**(21), 12156–12170 (2020)
3. Tang, B., Kramer, S., Fang, M.: A self-attention based message passing neural network for predicting molecular lipophilicity and aqueous solubility. J. Cheminform. **12**(15), 1–9 (2020)
4. Song, Y., Chen, J., Wang, W.: Double-head transformer neural network for molecular property prediction. J. Cheminform. **15**(27), 1–16 (2023)
5. Wu, Z., Ramsundar, B., Feinberg, E.N.: MoleculeNet: a benchmark for molecular machine learning. Chem. Sci. **9**(2), 513–530 (2018)
6. Koolman, J., Roehm, K.-H.: Color Atlas of Biochemistry. Thieme (2012)
7. Gómez-Bombarelli, R., Jennifer, N.W., Duvenaud, D.: Automatic chemical design using a data-driven continuous representation of molecules. ACS Cent. Sci. **4**(2), 268–276 (2018)
8. Bengio, Y., Goodfellow, I., Courville, A.: Deep Learning. MIT Press, 800 p. (2016)
9. Stokes, J.M., Swanson, K., Yang, K.: A deep learning approach to antibiotic discovery. Cell **180**(4), 475–483 (2020)
10. Ali, Y.A.: Hyperparameter search for machine learning algorithms for optimizing computational complexity. Processes **11**(2), 1–21 (2023)

1C Corporate Implementation Technology in ERP-System Deployment Projects in Russian and Localized Foreign Companies

Dmitry Yu. Stepanov(✉)

MIREA–Russian Technological University, Vernadskiy Avenue 78, 119454 Moscow, Russia
mail@stepanovd.com

Abstract. The paper contains analysis of 1C corporate implementation technology to deploy ERP-systems in Russian and localized foreign companies. The stages, activities and documents of the 1C methodology are described. Comparing spiral 1C methodology with cascade ADM model, issues of 1C CIT are identified: variability of stages, requirements collecting, user acceptance testing, data migration and cutover, which is important to be resolved for ERP-system implementation projects at Russian enterprises. Using the experience of deploying foreign software, it is proposed to adapt 1C CIT to ADM model, thereby making it relevant for use in localized foreign organizations with their own specifics.

Keywords: Corporate information systems · ERP-systems · 1C · corporate implementation technology · 1C CIT · ADM

1 Introduction

Leaving most foreign vendors from Russia in 2022 shifted the focus to local software solutions. The early dominance of foreign software products represented by SAP, Oracle, Microsoft and others is gradually coming to naught: many Russian companies, following import substitution and digital transformation strategies, are quickly transitioning to local software products [1]. In this regard, ERP-systems deserve special attention. This class of program systems, which belongs to the most representative automation standard, is introduced by various Russian solutions from 1C, Galaktika, Monolith and other software vendors [2, 3].

The ERP-systems from 1C company are the most popular nowadays [4]. Abandoning foreign programs and transitioning to 1C ERP solution is equivalent to the implementation project from scratch. 1C company has developed three methods of software deployment: standard implementation technology (1C SIT), fast result technology (1C FRT) and corporate implementation technology (1C CIT) [5]. However, for the middle and large enterprises and corresponding ERP-projects in which there is a high business uncertainty and employee resistance to change, the only one 1C CIT strategy is applicable.

V. Jordan et al. (Eds.): HPCST 2024, CCIS 2919, pp. 137–150, 2026.
https://doi.org/10.1007/978-3-032-20325-0_11

2 Problem Statement

The feature of the 1C CIT methodology is that it is aimed for implementing ERP-products for Russian enterprises, characterized by a low automation and, as a result, inexperienced end users. On the contrary, business processes of foreign companies are mostly automated, so their localization and transition from SAP, Oracle or Microsoft to 1C products have other difficulties. The purpose of this paper is to consider 1C corporate implementation technology for its application in ERP-projects in localized foreign organizations.

3 Waterfall, Iterative and Spiral Implementation Models

There are several classical software deployment models that are also suitable for the implementation of ERP-systems. The model defines basic principles of project work, ensures the reduction of typical project risks and delivery of software product in time. All existing implementation methodologies detail, expand and inherit activities from the three models:

- cascade;
- iterative;
- spiral,

as well as all possible hybrids of them [6]. ASAP, MDSS, OUP and ADM [7–10] are the examples of cascade-based implementation methodologies, iterative models are representable by Agile Scrum, Kanban and Feature Driven Development [11], while spiral schemes are demonstrated in Rapid Application Development, eXtreme Programming and 1C CIT [12]. Deployment of foreign ERP-systems is mainly performed under the cascade ADM methodology, for the implementation of 1C products of a similar software complexity the spiral 1C CIT approach is recommended. Let's focus on these two implementation methodologies.

3.1 ADM as a Cascade Implementation Model

For a long time, the ADM methodology based on the cascade implementation model has been used in both Russian and international ERP-projects, having virtually no alternative from domestic vendors [10]. The latter is due to ERP-systems are one of the most extensive automation class of software, the implementation of which is quite expensive and imposes special requirements on the organization and execution of the project. The ADM methodology has been used all over the world, which allowed it to be adapted to any specifics of customer or country. This methodology is well documented and contains a set of accelerators, that is, document templates to fill in during the project, as well as a description of activities and tasks to deliver the project content. The ADM model includes the following steps (Fig. 1):

- preparation, where project team is mobilized;
- analysis, here requirements are collected and their Fit/Gap-analysis is performed;
- design, where project documents are prepared and confirmed;

- built, software programs are developed and configured, as well as unit testing is carried out;
- test, integration and user acceptance tests are conducted;
- cutover, where productive system is prepared, master and transactional data are migrated into it, taking into account business constraints;
- hypercare support of productive operation and completing the project.

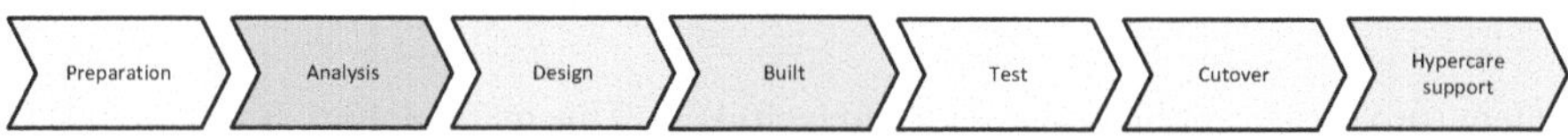

Fig. 1. Implementation stages according to the ADM methodology

The cascade ADM model distinguishes separate stages for analyzing requirements and conducting software testing, an unbroken sequence of stages, as well as no variability in the passing of stages.

3.2 Spiral Deployment Model by 1C

1C company recommends to implement complex software products using the 1C CIT methodology. This technology is based on a spiral deployment model with some elements from waterfall and iterative models [5]:

- project is divided into several implementation waves;
- within each wave a part of software product is realized using waterfall approach;
- built stage is conducted in iterations, in which the result is demonstrated to the customer according to the best practices of Agile.

The stages of 1C CIT includes:

- initialization;
- collecting requirements and their allocation over the waves of implementation,

then, for each wave of implementation, the following stages are repeated:

- design,
- built,
- preparation to pilot operation and/or pilot industrial operation;
- pilot operation and/or pilot industrial operation,
- preparation to productive operation;
- productive operation,

project closing is the final step (Fig. 2). The stages demonstrate a distinctive feature of this methodology: the variability of preparation to and execution of pilot operation and pilot industrial operation phases.

The 1C CIT methodology contains a detailed description of the stages and their activities, a list of documents, as well as examples and templates for some of them. 1C technology details questions related to project and change management, as well as

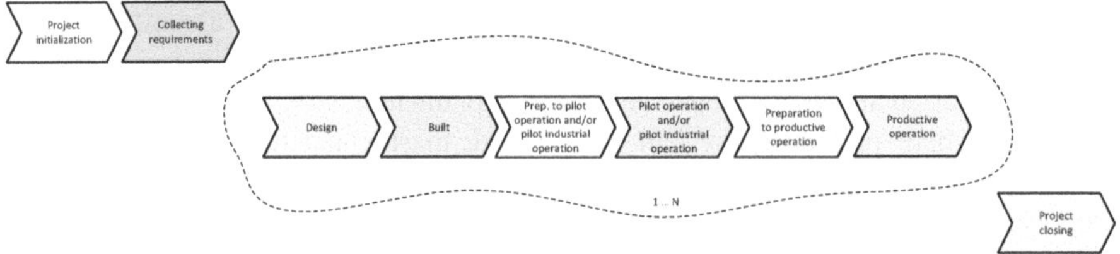

Fig. 2. Deployment stages according to the 1C corporate implementation technology

content (processes, applications, data and tech) [13]. This methodology is based on ISO 9000:2015 and ISO 9001:2015, as well as PMBoK [14]. There are key elements of this methodology given in tab. 1.

Table 1. Stages, key activites and documents of 1C CIT

Stage	Level	Activity	Document
Initialization	Project	Preparation of project charter	Project charter
	Project	Mobilization of project team	List of project participants
	Project	Organization of project office	Project office regulations
	Project	Kick-off meeting	Meeting minutes
	Project	Preparation of project document templates	Project document templates
Collecting requirements	Process, Application, Data, Tech	Interviewing customer's employees	Interview minutes
	Process, Application, Data, Tech	Preparation of pre-project report	Pre-project report
	Change	Creating the first version of organizational change plan	Organizational change plan
	Process, Application, Data, Tech	Shaping the document of functional and technical requirements	Functional and technical requirements

(*continued*)

Table 1. (*continued*)

Stage	Level	Activity	Document
Design	Tech	Deployment of a temporary infrastructure, development and testing environment for program software	Deployment meetings
	Process, Application, Data	Preparation of conseptual project	Conceptual project
	Application, Change, Data	Elaboration of concepts and techniques	Concepts of roles and authorizations, master data management, data migration, integration, user training, as well as testing programs and methods
	Process, Application, Data	Preparation of designs	Designs (methodological, technical etc.)
Built	Application	Iterations 1..N for the creation and unit testing of software system	Minutes of unit testing
	Application	Demonstrating developed and customized software system to the customer's users	Demonstration minutes
	Application	Documenting developments and customizations done	Specifications
	Process, Data	Creating test scenarios	Test scenarios
	Change	Key user training	Training minutes
	Application	Performing preliminary tests by contractor and customer's key users	Minutes of preliminary tests
Preparation to pilot operation and/or pilot industrial operation	Change	Preparation of cutover and business continuity plans	Cutover and business continuity plans

(*continued*)

Table 1. (*continued*)

Stage	Level	Activity	Document
	Application, Tech	Technical preparation of software system for pilot operation and/or pilot industrial operation	Minutes of technical preparation
	Data	Test data migration	Minutes of test data migration
	Change	Carrying out organizational chages at the customer	–
	Change	Key and end users training	Training minutes
Pilot operation and/or pilot industrial operation	Process, Application	Performing activites of pilot operation and/or pilot industrial operation by key/end users	–
	Application	Registration and elimination of defects	List of defects
Preparation to productive operation	Process, Application, Data, Tech	Updating the project documentation	Project documentation
	Tech	Technical preparation of software system for productive operation	Minutes of technical preparation
	Data	Productive master and transactional data migration	Minutes of productive data migration
	Change	End user training	Training minutes
	Application	User acceptance test	Minutes of acceptance test
Productive operation	Process, Application	Performing activites of productive operation by end users	–
	Application	Registration and processing of user requests	Lists of defects and user requests
Project closing	Project	Customer satisfaction assessment	Questionnaire of satisfaction with the project results

(*continued*)

Table 1. (*continued*)

Stage	Level	Activity	Document
	Project	Getting customer feedback	Customer feedback
	Project	Internal quality control of the project	–
	Project	The final meeting	Meeting minutes

4 ADM Methodology vs 1C Implementation Technology

The deployment stages of the ADM and 1C CIT methodologies can be visually represented by linearly, since both are based on the waterfall implementation model. This will allow them to be compared by the following parameters:

- stages;
- collecting requirements;
- user acceptance test;
- data migration;
- cutover and continuity plans.

4.1 Stages

The 1C CIT methodology introduces the stages of preparation to and execution of pilot operation and pilot industrial operation, which are absent in ADM. These stages are an emulation of the company's work for one of the past reporting periods in the 1C software system, which involves customer's key and end users. The difference between pilot industrial operation and pilot operation stages is that the results of emulation of the first one are further used in the industrial operation of software system. Both groups of stages are very different in meaning, content and goals from the test stage of ADM: the simulation results indicate the willingness of end users to work with the software system, but not readiness of the software product. The pilot industrial operation phase is similar to dual maintenance mode, when users repeat business operations both in an existing and is being implemented software systems. In the ADM methodology, such case is applicable only in pilot projects. The 1C implementation technology assumes the variability of pilot operation and pilot industrial operation stages, but there are no recommendations for their use (Fig. 3). Thus, the CIT method implies that each project team decides the presence and sequence of the stages by themselves.

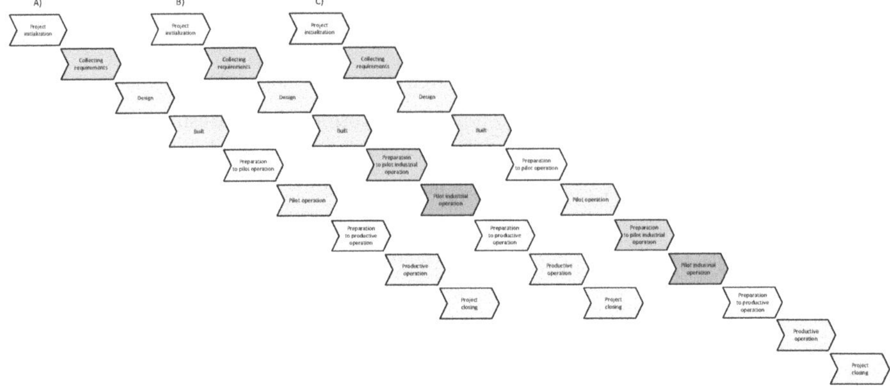

Fig. 3. Variability of stages in the 1C methodology: a) pilot operation stages; b) pilot industrial operation stages; c) both group of stages.

The sequence of stages in the 1C CIT methodology, containing the phases of preparation to and execution of pilot operation (Fig. 3A), seems to be more balanced. We will limit considering this variation of the 1C approach for the purpose of comparison with ADM. As you can be seen on Fig. 4, both methodologies are similar in terms of stages and their activities, with the exception of testing the software product. The testing stage from ADM is comparable with four phases from 1C CIT.

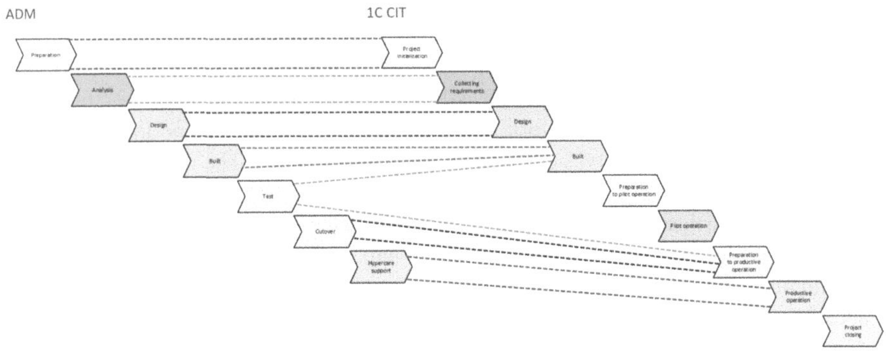

Fig. 4. Mapping ADM and 1C CIT stages

4.2 Collecting Requirements

The requirements collection strategy allows you to regulate how business needs will be identified, analyzed and prioritized [15]. According to ADM, the most effective way to collect business requirements in ERP-projects is to use a list of typical enterprise business processes from the relevant industry. The prepared business process map makes it possible to clarify specific needs of the client at the time of discussing operations of 4th level of detail contained in it (Fig. 5). The latter reduces losing business requirements, due to all processes and operations from the prepared map are discussed. The

1C methodology does not contain a strategy of collecting requirements. It is assumed that 1C demo database will be prepared in advance and used for further discussion of the requirements during its presentation to the customer. Although there is a register of business processes in the list of 1C CIT accelerators, most often it is prepared after identification of requirements.

4.3 User Acceptance Test

The cascade methodology ADM includes several types of software testing based on the V-model [16], in particular:

- unit test, which is carried out immediately after the completion of configuration and development of software product by technical specialists and developers;
- integration testing, which is conducted by technical specialists with subsequent demonstration of critical results to the customer;
- and finally, user acceptance test, in the form of end-to-end testing of business processes in all interconnected software systems. User acceptance testing is conducted based on pre-prepared test scenarios by key and end users of the client (Fig. 6A).

The first two types of testing from ADM have analogues in the 1C CIT method: a unit test and preliminary tests conducted during the built stage. The user acceptance test in the 1C methodology is mentioned only at the stage of preparation to productive operation, where the corresponding project document containing the results of all previously performed tests and emulations is signed-off. It also contains the results of additional verification of functional and non-functional requirements (Fig. 6B).

1st level	2nd level	3rd level	4th level
7. Sales	7.2. Operational sales	7.2.1. Sales orgstructure	7.2.1.1. Organizational structure of sales
7. Sales	7.2. Operational sales	7.2.2. Sales master data	7.2.2.1. Price list
7. Sales	7.2. Operational sales	7.2.2. Sales master data	7.2.2.2. Discounts
7. Sales	7.2. Operational sales	7.2.3. Sales to customer	7.2.3.1. Sales order creation
7. Sales	7.2. Operational sales	7.2.3. Sales to customer	7.2.3.2. Outbound delivery creation
7. Sales	7.2. Operational sales	7.2.3. Sales to customer	7.2.3.3. Print TORG-12
7. Sales	7.2. Operational sales	7.2.3. Sales to customer	7.2.3.4. Billing document creation
7. Sales	7.2. Operational sales	7.2.3. Sales to customer	7.2.3.5. Print Invoice

Fig. 5. An example of business process map for collecting user requirements according to the ADM methodology.

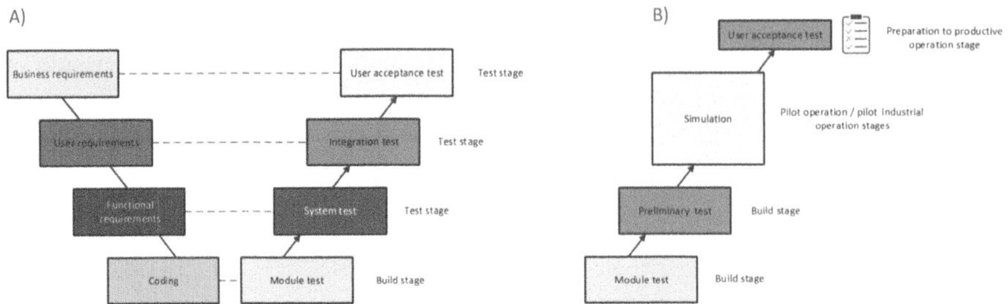

Fig. 6. Types of software testing in the methodology: a) ADM; b) 1C CIT

4.4 Data Migration

The data migration strategy implies a sequence of steps to migrate different kinds of data from legacy system to a new software solution [17]. According to ADM, following questions to be described in the concept:

- matrix or functional organization of migration team;
- maintaining a register of migration objects containing a list of all data objects, their migration methods, responsible for migration steps, the expected number of records for migration, etc.;
- the number of test waves of migration, allowing a rehearsal of data migration on a limited volume of test records;
- responsibility for conducting test waves of data migration between the contractor and customer;
- parameters to be controlled during test migrations, in particular, the execution time of all migration steps for each data object;
- preparation of data migration plan based on the register of migration objects, which contains deadlines and responsible for test waves and productive migration;
- early productive migration of master data and their double maintenance in software systems.

In the ADM methodology, test migration precedes any kinds of a software testing, it allows you to start checking migration tools and quality of migrated data in advance. Therefore, three test waves of data migration are introduced: one for the built stage and unit test, the other two for the test stage and integration, user acceptance testing (Fig. 7A). The 1C methodology has similar content, but it does not mention test waves of data migration. Moreover, in 1C CIT migration activities begin only with the stage of preparation to pilot operation and/or pilot industrial operation, while skipping built stage and testing carried out on it (Fig. 7B).

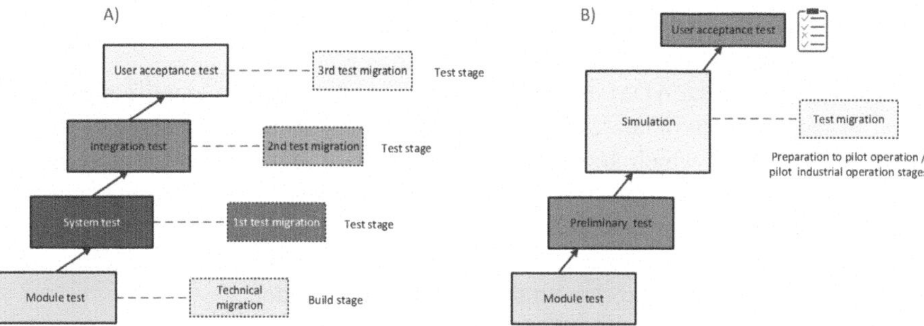

Fig. 7. Test waves of data migration preceding software testing in the methodology a) ADM; b) 1C CIT.

4.5 Cutover

Cutover activities are the most critical at the time of go-live in ADM. The tasks of transition period are detailed in cutover plan. The cutover plan combines and binds together the terms and duration of activities that are important for go-live of a software solution: business processes, data, applications and tech. Preparation of cutover plan for each operation of each business process requires:

- understanding what needs to be done to complete the operation in legacy system;
- analysis of business and technical prerequisites to be performed, as well as master and transactional data to be migrated in order to resume the execution of business operation in a new software system;
- comprehension if there is blackout period in the execution of operation when activity in legacy system is no longer conducted on it, but due to some limitations, it is not yet possible to start performing it in a new software solution (Fig. 8).

Usually, business continuity plan designed to ensure the smooth operation of organization in case of software crash is an integral part of business cutover plan. In general cutover plan combines both business and technical tasks, requires allocation of a separate team. Artifacts of the cutover and business continuity plans are mentioned once in the 1C CIT methodology but without any details or explanations.

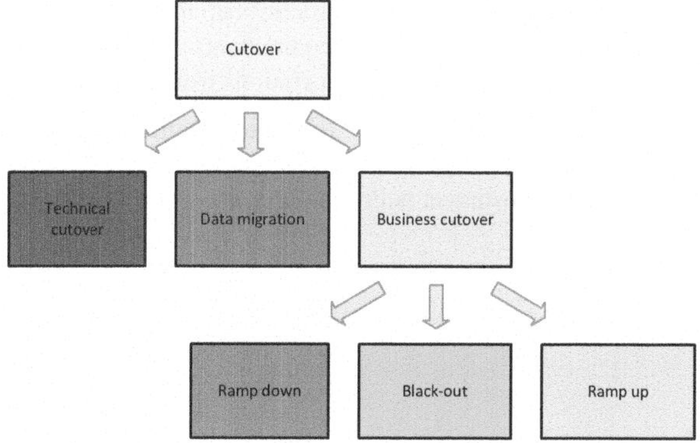

Fig. 8. Key activities of cutover

5 Points of Improvement for 1C Technology

Detailed analysis of the ADM and 1C CIT methods allowed us to identify their differences, which are given in tab. 2. The table also contains recommendations to improve the 1C methodology, which are applicable for software implementation projects both in Russian companies and localized foreign organizations. The proposed recommendations are aimed to improve the existing 1C CIT approach by correcting the identified drawbacks, as well as to expand the scope of 1C methodology usage by aligning some stages and content of their work with the cascade ADM model.

Table 2. Recommendations to improve the 1C CIT methodology

Activity	Disadvantage	Recommendation	Relevant for
Pilot operation and pilot industrial operation stages	No recommendations how to choose the stages	To add recommendations for the selection of project stages. For example, preparation to and execution of pilot operation stages are used in the projects of medium complexity, and the joint use of pilot operation and pilot industrial operation phases is allowed in highly complex and critical projects, where there are risks, that users will not be able to work with the new software system	Russian companies
Collecting requirements	Collecting requirements only by demonstrating a prototype of 1C system	To strengthen the approach by using a pre-prepared business process map to collect requirements within operations from it	Russian companies, localized foreign organizations
Data migration	There is no test data migration preceding software testing at built stage	To add test migration activity before unit and preliminary tests	Russian companies, localized foreign organizations
Cutover and continuity plans	No details	To add a detailed description of the plans to the methodology	Russian companies, localized foreign organizations

(continued)

Table 2. (*continued*)

Activity	Disadvantage	Recommendation	Relevant for
User acceptance test	User acceptance test is more meeting minutes than final stage of testing	To change content of the stages, aligning them with ADM, thereby shifting the focus from emulation to classical user acceptance testing	Localized foreign organizations

6 Conclusions

We discussed two popular models for the implementation of corporate information systems within the paper: the cascade ADM methodology, used in large foreign deployment projects, and the spiral 1C CIT approach, used for the implementation of software in Russian companies. The differences between these methodologies are revealed. The analysis identified 1C CIT areas that require special attention: variability of stages, requirements collecting, user acceptance testing, data migration, as well as cutover. Recommendations are given to make the 1C methodology better, allowing to improve its use in the implementation projects at Russian companies, as well as to expand using it for ERP-projects in localized foreign organizations, having a high level of automation.

References

1. Katasonova, N.S.: The big bang method and franchise strategy for implementation and import substitution of ERP-systems. Corp. Inf. Syst. **23**(3), 1–11 (2023). https://corpinfosys.ru/archive/2023/issue-23/227-2023-23-implementationstrategy. (in Russian)
2. Petrov, S.V.: Overview of Russian software for import substitution of MES, ERP2 and BI corporate information systems (part 1). Corp. Inf. Syst. **20**(4), 31–41 (2022). https://corpinfosys.ru/archive/issue-20/208-2022-20-applicationsoverview. (in Russian)
3. Petrov, S.V.: Overview of Russian software for import substitution of MES, ERP2 and BI corporate information systems (part 2). Corp. Inf. Syst. **21**(1), 38–48 (2023). https://corpinfosys.ru/archive/issue-21/218-2023-21-applicationsoverview. (in Russian)
4. Bobrovnikov, A.E.: Introduction to ERP-System Implementation Project Management. 1C-Publishing, Moscow (2021). (in Russian)
5. Zimin, K.: Corporate implementation technology. Inf. Manage. **3**, 77–84 (2013). (in Russian)
6. Habadi, A.: An Introduction to ERP-systems: architecture, implementation and impacts. Int. J. Comput. Appl. **167**(9), 1–4 (2017)
7. Sullivan, G.: SAP: Project Management and Implementation Guide. SAP Press (2014)
8. Blokdyk, G.: Oracle Unified Method. 5STARCooks (2018)
9. Shankar, C., Bellefroid, V.: Microsoft Dynamics Sure Step 2010. Packt Publishing (2011)
10. Stepanov, D.Yu.: The lifecycle of ERP-system implementation projects by the example of SAP and 1C package solutions and custom developments. Habr, Axenix official page. https://habr.com/ru/companies/axenix/articles/799333/. (in Russian)

11. Cohn, M.: Agile Estimating and Planning. Pearson Publisher (2005)
12. Stellman, A., Greene, J.: Learning Agile: Understanding Scrum, XP, Lean and Kanban. O'Reilly Media (2013)
13. Stepanov, D.Yu.: Analysis, design and development of corporate information systems: theory and practices. Russ. Technol. J. **8**(3), 227–238 (2015). (in Russian)
14. Shirenbek, H., Lister, M., Kirmse, Sh.: A Guide to the Project Management Body of Knowledge, 6th edn. PMI Publisher (2017)
15. Terentiev, I.M.: Requirement analysis in ERP-system implementation projects. Corp. Inf. Syst. **2**, 23–29 (2018). https://corpinfosys.ru/archive/issue-2/139-2018-2-analysisstrategy. (in Russian)
16. Terentiev, I.M.: Testing strategy in ERP-system implementation projects. Corp. Inf. Syst. **3**, 39–45 (2018). https://corpinfosys.ru/archive/issue-3/141-2018-3-testingstrategy. (in Russian)
17. Eremenko, Y.O.: Features of data migration in SAP ERP. Corp. Inf. Syst. **7**(3), 22–28 (2019). https://corpinfosys.ru/archive/2019/issue-7/67-2019-7-migration. (in Russian)

Algorithms for Efficient Aspect Analysis of Scientific Publications Based on Semantic Graph Clustering

Elena Degtiareva(✉), Elena Kozlova, and Elena Kryuchkova

Polzunov Altai State Technical University, Barnaul, Russia
vopilova.elena@gmail.com

Abstract. The paper presents a semantic model of scientific thesaurus represented in the form of the domain semantic graph. Thematic scientific encyclopedias we use as a source of training data for model construction. Semantic graph clustering algorithms built on the basis of a thesaurus are proposed as a tool for aspect-oriented analysis of a scientific publication and analysis of a scientific field as a whole. The domain semantic graph «Mathematics» was constructed as a result of processing the text of the mathematical encyclopedia in five volumes. Examples of scientific publication analyzing experiments are presented. The task of dynamic aspect analysis is discussed. The proposed algorithm for aspect analysis of a scientific text is based on the transformation of the domain semantic graph into the publication semantic graph. The problem of automatic grouping scientific terms into groups corresponding to the different science branches is posed. The experiment results of domain graph clustering are presented.

Keywords: Semantic Graph · Text Processing · Unstructured Texts · Clustering

1 Introduction

Scientific text processing has special features. Automatic analysis of scientific texts includes wide variety of tasks, such as the domain terminology extraction, creation of scientific terms thesaurus [1, 2], abstracting and annotating [3], information retrieval [4]. Scientific texts have some syntactic and semantic properties, which should be taken into account in the designing of algorithms for automatic processing. Highly specialized terminology and specifics of this terminology using in scientific publications significantly influence on a domain semantic model. The scientific vocabulary contains multiword terms, moreover, multiword terms constitute the majority of using terminology in many subject domains. However, the word order in a term may be different in Russian, and it is possible to use a truncated form of a term. It is required to disambiguate the term and to execute lemmatization of each individual word. Also important is the fact that scientific texts contain pattern word sequences, which are not relevant to the domain vocabulary, e.g. «what was required to be proved», «it is obvious that», etc.

Currently, there are methods for automatic scientific text analysis, some of them use a prepared domain-specific semantic core, for example, it can be a domain formal

V. Jordan et al. (Eds.): HPCST 2024, CCIS 2919, pp. 151–162, 2026.
https://doi.org/10.1007/978-3-032-20325-0_12

ontology [5]. Domain knowledge can be represented as a graph [6, 7], these approaches make it possible to use graph theory methods and algorithms.

A fair amount of research in computational linguistics is based on probabilistic-statistical methods of text processing [8, 9]. Substantial increase of computational performance and the propagation of parallel programming cause the common using of neural network models, large language models (LLM) [11].

This paper proposes a semantic model of a scientific domain thesaurus, automatically built as a result of text processing of a specialized encyclopedia. Completed model uses for aspect-oriented analysis of a scientific text, proposed algorithms are based on semantic graph clustering. Experiments were conducted for aspect-oriented analysis of scientific publications and construction the structure of the corresponding science as a whole. The algorithms for aspect analyzing of scientific texts and determining the aspect transition dynamics are considered.

2 Scientific Domain Semantic Graph

Scientific terminology has properties, in our opinion, the dominating one is the multiword composition of scientific terms. The meaning of the term is defined by the strong relations between its lexemes (individual words), therefore, word relations describe the text semantics, in which this term appears. For example, occurrence frequency analysis of individual words «number» or «normal» shows us nothing about the subject theme of a text, since these words may occur in texts from different mathematic sections, or from other subjects, or even from commonly used topics. However, multiword terms «law of large numbers» and «normal distribution» clearly indicates belonging of the text to Probability theory, at the same time the term «normal number» or «large prime number» in the text indicates that this text belongs to Number theory. Moreover, each individual word in a term can be highly specialized («hypotenuse», «separable», «theorem», etc.) or pertained to the common vocabulary («field», «root», «plain», etc.). But the principle does not change: the uniqueness of a term is determined by the set of its lexemes.

The uniting of independently extracted lexemes from texts into a merged term is a difficult task [12], which can be solved by using a prepared list of terms from scientific encyclopedias. The task of domain term extracting can be accomplished by extracting knowledge from corresponding scientific domain encyclopedias. The text of scientific encyclopedia is partial structuring, extensively covers professional terminology, has an extended descriptions of terms. These facts make it possible to automatic extraction of subject terminology and some semantic relations between terms.

As a knowledge representation model of a scientific domain we will use a weighted oriented multigraph $G_{domain} = (V, E)$, whose vertices V correspond to domain terminology and arcs E represent semantic relations between them. The vertices of the graph contain scientific terms divided into two types:

- *Basic terms* correspond to the titles of individual thematic encyclopedia papers. Basic terms and relations between them form the semantic core of the scientific field.

- *Fixed word collocations* of the subject area are syntactically coherent n-grams, extracted from texts of the encyclopedia papers. In the future some fixed word collocations can change their status and become basic terms. Fixed word collocations facilitate in solving tasks of paper text analyzing.

Semantic relations of different types are set up between scientific terms: generalizations, definitions, associations. The type of relation is determined by the structural block of the subject encyclopedia paper, from which the term was extracted. In the semantic graph G_{domain}, the weight of each arc $e=(v_1,v_2)$, $e \in E$ determines the semantic significance of vertex v_2 for the corresponding term v_1. The arc weights of the graph G_{domain} are normalized to the range [0;1].

Exactly associations are primary in human text comprehension. The algorithm for computing of the arc weights in the domain semantic graph and a comparative analysis of calculation methods are beyond the scope of this paper and are discussed in [13]. In this paper, a formula for calculating the significance of an associative relation is

$$Z(X, T) = \log_2 N \cdot \sum_{i=1}^{n} TF(w_i, T) \cdot IDF(w_i)$$

where w_i – an individual word from the set of encyclopedia words;
$X = \{w_1, w_2, \cdots, w_n\}$ – n-gram, which is the set of the subject encyclopedia words;
T – n-gram, which is the basic term of the subject area;
$Z(X, T)$ – significance of n-gram X for the term T;
$TF(w_i, T)$ – frequency of the word w_i in the encyclopedia paper, describing the term T;
$IDF(w_i)$ – inversion of the frequency of the word w_i in all encyclopedia papers.

Presented in this paper the computational experiment results are based on terminology, automatically extracted from the five-volume Mathematical Encyclopedia [14], which contains more than 6.5 thousand papers about mathematical terms.

3 Scientific Publication Semantic Graph

In the reading process, a human uses accumulated knowledge about the subject area to understand any text, scientific text especially. To increase interpretability of reading text, a human's brain combinates and overlays information from the reading text and accumulated knowledge about the subject. In our model, we used such principles. In this paper we propose using the graph G_{domain} as a text fragment context, in the text processing the extracted information from the analyzed text applies to the domain graph G_{domain}. This approach takes into account the fact that some terms can't be explicitly mentioned in the publication text, but these terms can be semantically related to the content of analyzed scientific publication. However, such terms are included in the general semantic context of the analyzed publication due to the associations that exist in our minds. In analyzing of the scientific text the domain semantic graph G_{domain} is transformed into the scientific publication semantic graph G_{text}, which contains the term usage statistics in the text in addition to domain-specific semantic relations.

In the building the scientific publication graph G_{text}, we calculate the vertex weights of the graph G_{domain}, then we propagate the calculated vertex weights to the neighboring nodes, connected with them by semantic relations. Propagation of the source vertex weight depends on spatial attenuation: the most distant neighboring vertices get a smaller additional weight.

The final weight of the vertex T_i with the additional weight from vertex T is calculated as

$$f(T_i) = f_{init}(T_i) + f_{init}(T_i) \cdot p(T, T_i) \cdot \gamma^{l(T,T_i)}, \quad (1)$$

where $f_{init}(T_i)$ – appearance frequency of the term T_i in the analyzed text,
γ – spatial attenuation coefficient,
$l(T, T_i)$ – count of arcs on the shortest path,
$p(T, T_i)$– product of arc weights on this path from T to T_i.

As a rule, a scientific publication text has following structure: first, related issues are briefly overviewed, then the main problems are discussed, the final part contains a brief list of the work conclusions. Moreover, even the main part of a scientific article follows the logic of the material presentation, thematic aspects are changing from one to another. The analysis algorithm should take into account the fact that each text fragment uses the context of the previous text fragment.

For tracing the transition dynamics from one thematic aspect to another we introduce the temporal attenuation coefficient δ. In the processing of each text fragment, we use vertex weights of the graph G_{text}, which is calculated by formula (1), with temporal attenuation:

$$a(T_i) = f(T_i) \cdot \delta, \quad (2)$$

which realizes the significance reduction of the information from previous text fragment. We can change contribution of the preceding text to analysis of the current fragment by varying the temporal attenuation coefficient in the range [0;1].

In the processing of a publication the graph G_{text} modifies: new vertices can be added and their weights increase according to formula (1) some vertices can be deleted after decreasing their weights according to (2) when $\delta < 1$.

4 Cluster Analysis of a Scientific Publication

Clustering of the scientific paper graph G_{text} can be a tool for determining the subject of the text of the scientific publication. Any scientific publication contains a list of author's keywords, which makes it possible to consider a publication text as partially marked up. The sufficiency of the proposed model was estimated by comparing the detected thematic clusters with the list of publication author's keywords. The results of the text processing experiments [15, 16] are summarized and the quantitative characteristics for each keyword length are presented in Table 1.

Table 1. Keyword definition recall depending on the keyword length

Keyword length	Count of author's keywords	Count of occurrences in publication texts	Definition recall
1	16	2–47	0.25
2	123	1–66	0.87
3	31	1–24	0.97
4	13	1–10	1.0
5	3	2–13	1.0

If we ignore temporal attenuation, cluster analysis makes it possible to detect the entire range of publication topics. And conversely, for analyzing the dynamics of content presentation and transitions between topics we use temporal attenuation coefficient. An example of analyzing the transition dynamics of text thematic accents is shown in Figure 1.

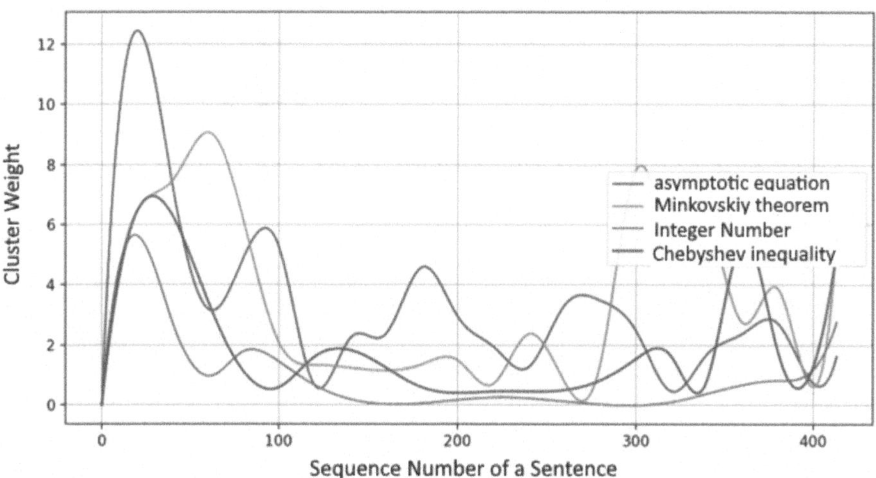

Fig. 1. Dynamics of cluster weight changes in the paper «Asymptotic decomposition and asymptotic formula for the root of the transcendental equation with a parameter».

5 Domain Graph Cluster Analysis

The goal of the clustering of a semantic domain multigraph is grouping a set of terms into topics, such that each cluster represents some scientific section, and the content of the cluster consists of terms related to that section. The using arc weights, corresponding to the significance of semantic relations, makes it possible to extract subgraphs with strong semantic relations in the domain graph G_{domain}. Detected clusters of the graph G_{domain} can be matched to scientific sections, after that we can set a science section label to each term.

Clustering is a milestone stage in the construction of the domain semantic core G_{domain}. As a result, each term will get a science section label, which will be use for the correspondence analysis of texts to some science section.

Vertices in the graph G_{domain} have no weights, instead of then we use arc weights for graph clustering. The detected term clusters must have strong internal relations and weak external relations between clusters. Before clustering the graph G_{domain}, we constructed the maximum cost spanning tree G_{tree} by removing N edges with minimum weight. After removing of the edges from the spanning tree G_{tree} the vertices of the graph G_{domain} divided into $N+1$ independent clusters. The algorithm is based on modification rules that have to be applied to an edge. Only three deleting rules might be applied: cut light edge from the graph without restoring, cut it with restoring if a result of modification rule is a small cluster, and append an edge to the cluster of its immediate neighbors.

To choose a clustering algorithm for the spanning tree G_{tree}, the analyzing their structure is necessary, including the distribution of weak and strong relations between vertices in the original domain graph G_{domain} and in the constructed spanning tree G_{tree}. The results of structure analyzing of graphs G_{domain} and G_{tree} are shown in Figure 2.

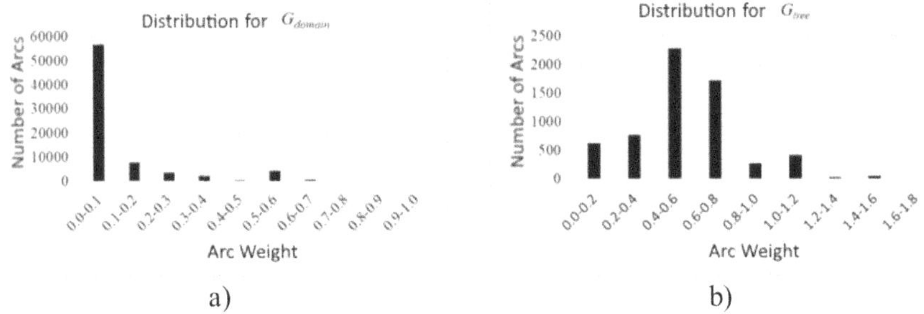

Fig. 2. The structure of graphs a) G_{domain} and b) G_{tree}

According to the UDC classifier, the Mathematics can be divided into 40-50 sections, such values of the count of clusters that were taken as a basis. As a result of experiments, it is determined that the required count of clusters can be extracted only if the minimum count of relations in the cluster is equal to 7 edges. The sizes of detected clusters range widely; this fact corresponds to the real structure of the Mathematics domain: there are large scientific sections, which are full of terms («Computational Mathematics» or «Differential Calculus»), and more specialized sections («Graph Theory» or «Combinatorics»). Examples of detected terminological clusters are shown in Table 2.

Table 2. Examples of detected clusters

Cluster	Count of terms in the cluster	Examples of terms in the cluster
Mathematical logic	229	Normal conjunctive form Logic-mathematical calculus Natural logical conclusion Intuitionistic calculus of propositions Law of double negation Implication
Number theory	245	Fractional distribution Arabic numeral Arithmetic progression Infinite decimal fraction Diagonal continued fraction Golden ratio Denominator Multiple
Random process	241	Random sequence Probabilistic process Prior probability Brownian motion process Bernoulli test Geometric probability Experiment planning Convergence in probability Correlation coefficient Reliability Random variable transformation

By counting the term frequency in the publication S we can compute weights of those sections in the graph G_{domain}, whose terms are used in the publication S, and the relative frequency $p_{section}(X,S)$ of each section X:

$$p_{section}(X, S) = \frac{V_{section}(X, S)}{W(S)},$$

where $V_{section}(X, S)$ – total frequency of terms of the publication S, owned by the section X,

$W(S)$ – the total frequency of terms of the text S, contained in all sections of the graph G_{domain}.

$$W(S) = \sum_{i=1}^{n} V_{section}(X_i, S).$$

By comparing the relative frequencies $p_{section}(X, S_1)$ and $p_{section}(X, S_2)$ of texts S_1 and S_2 we can estimate the belonging of these texts to the same scientific sections. Moreover, we can detect that the theme of one text is partly contained in the broader theme of another text. Experiments were conducted to analyze the correspondence of the publication texts to the scientific specialties. The inclusion coefficient $k_{incl}(S, S_b)$ characterizes the degree of conformity that the publication text S corresponds to the thematic aspects of the reference text S_b:

$$k_{incl}(S, S_b) = \frac{\widetilde{W_c}(S)}{W(S)}$$

where $\widetilde{W_c}(S)$ – sum of frequencies of terms in the text S contained in the clusters of the reference text S_b.

We used the most common approach to train the scientific field classifier by using field-oriented unstructured texts. As a reference text S_b describing the scientific field, we have chosen the scientific specialties, approved by the Ministry of Science and Higher Education of the Russian Federation. Examples of publication analysis are shown in Table 3.

Table 3. Correspondence analysis of the paper texts to the scientific specialty

Paper title	Most relevant scientific specialty	Cluster centers	k_{incl}, %
Proper cyclic symmetries of multidimensional continued fractions	1.1.1. Real, complex and functional analysis	1. Number theory 2. Holomorphic function 3. Algebraic group 4. Banach space 5. Differentiation of a function 6. Real-valued function 7. Continuous map 8. Variational calculus 9. Group presentation 10. Riemann surface 11. Linear system	84
	1.1.5 Mathematical logic, algebra, number theory and discrete mathematics	1. Number theory 2. Galois group 3. Mathematical logic 4. Abelian group 5. Simple semigroup 6. Group presentation	59

(*continued*)

Table 3. (*continued*)

Paper title	Most relevant scientific specialty	Cluster centers	k_{incl}, %
Coincidence of set functions in quasiconformal analysis	1.1.1 Real, complex and functional analysis	1. Continuous map 2. Banach space 3. Real-valued function 4. Integral equation 5. Differentiation of a function 6. Number theory 7. Riemann surface 8. Linear system 9. Holomorphic function 10. Group presentation	62
	1.1.5 Mathematical logic, algebra, number theory and discrete mathematics	1. Set of functions 2. Mathematical logic 3. Number theory 4. Convex polyhedron 5. Abelian group 6. Galois group 7. Recursive function 8. Group presentation 9. Ring modulus	33
A circle criterion for a generalized cross graph in terms of minimal excluded minors	1.1.5 Mathematical logic, algebra, number theory and discrete mathematics	1. Convex polyhedron 2. Abelian group 3. Set of functions 4. Recursive function 5. Mathematical logic 6. Number theory	76
Minimization of representations of the logical function in schaeffer and pierce bases	1.1.5 Mathematical logic, algebra, number theory and discrete mathematics	1. Mathematical logic 2. Ring modulus 3. Number theory 4. Recursive function 5. Galois group 6. Group presentation	89
Classification of periodic differential equations by degrees of non-roughniss	1.1.2 Differential equations and mathematical physics	1. Variational calculus 2. Linear system 3. Banach space 4. Vector field 5. Number theory	76

(*continued*)

Table 3. (*continued*)

Paper title	Most relevant scientific specialty	Cluster centers	k_{incl}, %
	1.1.1 Real, complex and functional analysis	1. Variational calculus 2. Linear system 3. Banach space 4. Number theory 5. Group presentation 6. Differentiation of a function	72
Linear differential holding game with a break	1.1.7 Theoretical mechanics, machine dynamics	1. Recursive function 2. Variational calculus 3. Linear system	80
	1.1.5 Mathematical logic, algebra, number theory and discrete mathematics	1. Recursive function 2. Set of functions 3. Mathematical logic	75
Thermal conductivity in a homogeneous strip with a linear change in thickness under boundary conditions of the first kind	1.1.2 Differential equations and mathematical physics	1. Vector field 2. Integral equation 3. Variational calculus 4. Linear system	66
	1.1.1 Real, complex and functional analysis	1. Integral equation 2. Integral transformation 3. Riemann surface 4. Variational calculus 5. Real-valued function 6. Group presentation 7. Linear system	49
Neumann boundary condition for a nonlocal biharmonic equation	1.1.1 Real, complex and functional analysis	1. Integral equation 2. Real-valued function 3. Linear system 4. Variational calculus 5. Banach space 6. Algebraic group 7. Differentiation of a function	93
	1.1.2 Differential equations and mathematical physics	1. Integral equation 2. Linear system 3. Variational calculus 4. Banach space	68
	1.1.9 Fluid, gas and plasma mechanics	1. Integral equation 2. Linear system 3. Variational calculus 4. Galois group	68

6 Conclusions

In this paper we have proposed and evaluated two semantic graphs exploiting information both from the science general context in G_{domain} and from a scientific publication in G_{text}. We have shown that their combination improve the results of analysis. We have proposed the clustering algorithm for a weighted semantic graph based on the selective cutting off the minimum weighted arcs of the spanning tree, constructed for the graph. The efficiency of the clustering algorithm and the qualitative composition of the detected thematic clusters have been discussed. The issues of the scientific publication aspect analysis have been considered, and the method for determining the correspondence of the publication content to the scientific specialty have been proposed. The proposed algorithms have been developed in the Python programming language using the Natasha library [17], which solves such basic natural Russian language processing tasks as tokenization, sentence extraction, lemmatization, morphological and syntactic analysis.

References

1. Bruches, E.P., Batura, T.V.: Method for automatic term extraction from scientific articles based on weak supervision. Vestnik NSU Ser. Inf. Technol. **19**(2), 5–16 (2021)
2. Morozov, D.A., Glazkova, A.V., Tyutyulnikov, M.A., Iomdin, B.L.: Keyphrase generation for abstracts of the russian-language scientific articles. Vestnik NSU Ser. Linguist. Intercultural Commun. **21**(1), 54–66 (2023)
3. Altmami, N., Menai, M.: Automatic summarization of scientific articles: a survey. J. King Saud Univ. Comput. Inf. Sci. **34**, 1011–1028 (2020)
4. Benites, F.: Information retrieval and knowledge extraction for academic writing. Digit. Writ. Technol. High. Educ., 303–315 (2023)
5. Borovikova, O.I., Kononenko, I.S., Sidorova, E.A.: An approach to information extraction from clinical trials protocols on the basis of medical ontology. Syst. Inf. **9**, 93–110 (2017)
6. Beliga, S., Mestrovic, A., Martincic-Ipsic, S.: An overview of graph-based keyword extraction methods and approaches. J. Inf. Organ. Sci. **39**, 1–2 (2015)
7. Lunev, K.V.: Graph methods for computing semantic similarity of a pair of keywords and their application to the problem of keywords clustering. Softw. Eng. **9**(6), 262–271 (2018)
8. Dubinina, E.Y.: Automatic extraction of key lexical units of the scientific texts at the process of summarization. Proc. SUAP Sci. Session **3**, 115–118 (2018)
9. Hossari, M., Dev, S., Kelleher, J.D.: TEST: a terminology extraction system for technology related terms. In: Proceedings of the 2019 11th International Conference on Computer and Automation Engineering, pp. 78–81 (2019)
10. Danilov, G., Ishankulov, T., Kotik, K., Orlov, Yu., Shifrin, M., Potapov, A.: The classification of short scientific texts using pretrained BERT model. Public Health Inform. **281**, 83–87 (2021)
11. Dunn, A., et al.: Structured information extraction from complex scientific text with fine-tuned large language models. arXiv (2022)
12. Papagiannopoulou, E., Tsoumakas, G.: A Review of keyphrase extraction. CoRR (2019)
13. Vopilova, E.V.: Characteristic functions for calculating the significance of terms in a semantic model of scientific knowledge representation. In: Proceedings of the 9th International Conference Knowledge - Ontologies - Theories (KONT-2023), pp. 49–53 (2023)
14. Vinogradov, M.: Encyclopedia of Mathematics. The Soviet Encyclopedia (1977)

15. Zagrebina, S.A. (ed.) Bulletin of the South Ural State University, Series Mathematics. Mechanics. Physics **14 (2-4)** (2022)
16. Sbornik: Mathematics. Sci. J., 9-12 (2019)
17. Natasha: Tools for Russian NLP: segmentation, embeddings, morphology, lemmatization, syntax, NER, fact extraction. https://github.com/natasha

Model for Forming a National University Ranking System in Kyrgyzstan

Bibigul Koshoeva(✉), Aigiuzel Bakalova, Bekzhan Torobekov, and Alima Azimova

I. Razzakov Kyrgyz State Technical University, Aitmatov Avenue 66, 720044 Bishkek, Kyrgyz Republic
koshoeva@kstu.kg

Abstract. The article examines the formation of a national university ranking system in the Kyrgyz Republic (KR). It proposes a university ranking management business process model, an automated university ranking system model presented in the form of an IDEF0 diagram and a data flow diagram, and an architectural solution for the automated university ranking system. Key evaluation criteria and assessment methodologies are analyzed in accordance with international standards. The automation of data analysis based on information technology will enhance the transparency, objectivity, and efficiency of the ranking system.

Keywords: University Ranking · Business Process Model · Automated System Model · MVC Architecture · Information Technology

1 Introduction

The modern higher education system is evolving under the conditions of globalization and increasing competition for academic and material resources. Universities strive to enhance their attractiveness by recruiting leading researchers, qualified faculty members, and promising students. In this context, the development of university rankings plays a crucial role, as they provide an objective assessment of institutional performance and effectiveness. However, in Kyrgyzstan, there is no unified methodology for evaluating universities, which hinders objective comparison and the strategic development of higher education.

The aim of this study is to develop a business process model for managing university rankings and a model of an integrated automated university ranking system for the Kyrgyz Republic.

The work is based on statistical data, analytical reports and scientific articles, uses systematic, statistical and comparative approaches. The KR Education Development Program for 2021–2040 provides for the creation of the National model of HEIs rating to prepare for international ratings.

The objectives of the study include analyzing existing rating systems, selecting indicators and data collection methods, designing an information system and conducting

V. Jordan et al. (Eds.): HPCST 2024, CCIS 2919, pp. 163–174, 2026.
https://doi.org/10.1007/978-3-032-20325-0_13

experimental research to determine the rating of higher education institutions of the Kyrgyz Republic [1–5].

The analysis of global rating systems and systems of the neighboring countries has shown that the introduction of the Bologna Process principles in higher education in Kyrgyzstan has led to the emergence of new criteria that determine the quality of educational services, their role and recognizability in society, as well as the position of HEIs at different levels [6, 7].

Rating of HEIs has become an important tool for evaluating their performance. The results of well-known rankings for 2022, including ARWU, QS and Times Higher Education, as well as the Spanish Webometrics ranking, were reviewed. It was revealed that the achievement of the rating targets contributes to the improvement of scientific and innovative activities of universities and the increase in revenues from them [1, 5].

2 Research Methodology

The studies conducted by Z. Karazakova and N. Sheranova confirm the necessity of internal evaluation systems within universities in Kyrgyzstan; however, their insufficient level of digitalization complicates the automation of related processes [1–3]. Y. Zhou and N. Asipov emphasize the impact of ranking mechanisms on the quality of education and national development, highlighting the importance of ICT and international standards [2, 3]. Within the framework of this research, leading global and regional ranking systems were analyzed, including the criteria used for their formation and the principles of their integration into a national model.

Modern higher education is characterized by increased competition for resources and academic positions. Universities strive to attract the best specialists, which makes the development of rating systems a key element in improving the quality of education and its compliance with international standards.

The purpose of the national rating is to create an independent system of evaluation of universities and educational programs according to the criteria of quality and internationalization. It will contribute to ensuring high quality of knowledge and competencies of graduates.

The objectives of the national rating of HEIs are as follows:

1. providing objective information on the status and dynamics of HEIs in order to increase their prestige and competitiveness.
2. identification of potential and problems to improve the efficiency of HEIs in management decisions.
3. conducting a systematic evaluation of HEIs' performance.
4. measuring various aspects of HEIs' performance and providing information on educational services.
5. developing methods for evaluating HEIs and educational programs.

The methodology of forming a rating system for evaluating the educational activity of HEIs has been developed, including:

- System of indicators and indicators for rating.
- Methods of data collection.

- Business process models (AS-IS and TO-BE).
- Conceptual and functional models of the rating system, including the infologic model.

The formation of university rankings involves the justification and selection of a representative number of indicators and evaluation criteria for university performance, each assigned appropriate weighting values. These form the basis of the experimental research database [1, 5]. The collection and processing of data, along with the calculation of scores according to the methodology, is a complex and multi-faceted process that requires significant time, material, and human resources. In this context, automating the university ranking process through information technologies becomes the most rational and effective approach for decision-making [3].

An integrated rating system is a software and methodological complex that provides collection, verification, processing, analysis and presentation of data on the performance of assessment objects on the basis of unified criteria and indicators.

The development of the ranking system model included the following stages:

1. Identification of key university evaluation criteria [1, 4].
2. Development of a ranking management business process model (IDEF0).
3. Creation of a logical-structural model of the automated system (DFD).
4. Development of the system architecture based on the MVC pattern.
5. Development of a methodology for data collection and processing [1, 5, 6].
6. Implementation and pilot testing of the automated system [6, 7].

3 Case Study for University Ranking in Kyrgyzstan

To understand how to implement an institutional university ranking system while taking into account all evaluation criteria, it was decided to develop a university ranking management business process model in the form of an IDEF0 diagram (Fig. 1):

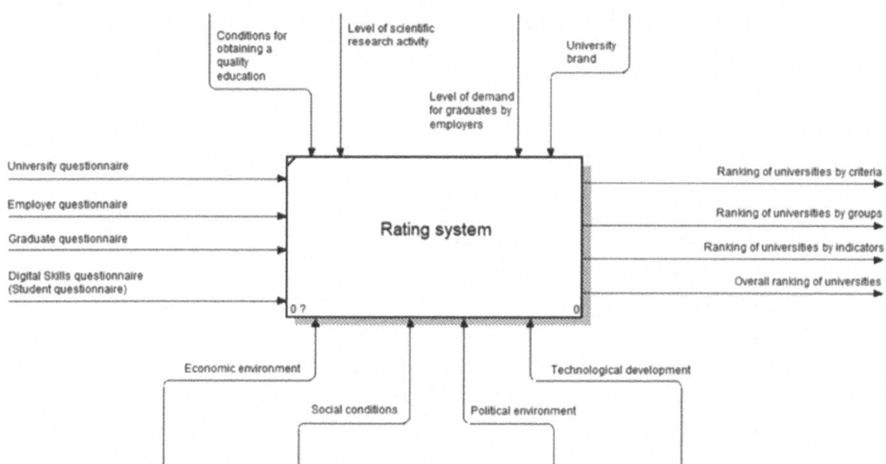

Fig. 1. Business process model of university ranking management system

which can also be represented in the form of functionality:

$$y = F(x, u, f) \tag{1}$$

where x – input parameters, $\overrightarrow{x}(t) = [x_1, x_2, x_3, x_4]$ – questionnaires,
where x_1 – university questionnaire,
x_2 – employer questionnaire,
x_3 – graduate questionnaire,
x_4 – digital skills questionnaire (student questionnaire);

y – output parameters, $\vec{y}(t) = [y_1, y_2, y_3, y_4]$,

where y_1 – rating result by levels (criteria),
y_2 – by groups,
y_3 – by indicators and.
y_4 – overall ranking of universities;

u – management, $\overrightarrow{u}(t) = [u_1, u_2, u_3, u_4]$ – rating criteria,

where u_1 – conditions for obtaining quality education,
u_2 – level of research work,
u_3 – the level of demand for graduates by employers and.
u_4 – university brand;

f – mechanisms (disturbing effect), $\overrightarrow{f}(t) = [f_1, f_2, f_3, f_4]$,

where f_1 – economic environment,
f_2 – social conditions,
f_3 – political environment,
f_4 – technological development.

During the research work, 4 criteria for forming the ranking of universities in the Kyrgyz Republic were developed:

- Conditions for ensuring quality education;
- Level of research activity;
- Level of demand for graduates by employers;
- University brand,

divided into 12 groups, each of which provides for several assessment indicators (integrated factors) [4–7].

The developed university ranking methodology is based on international standards. The evaluation criteria align with the principles of objectivity (using external data or verified university information), verifiability (quantitative indicators), and uniformity (based on statistical reporting to ensure accurate comparison).

To test the methodology and determine the university rankings, a data collection, processing, and analysis (evaluation) procedure was developed. The primary data on university performance is collected through the completion of questionnaire forms. The methodology for constructing the questionnaires may vary depending on the specific goals and context in which the questionnaire is used. Several steps were identified that can be included in the methodology:

1. **Defining the Purpose of the Questionnaire:** Determine what you want to learn or measure with the questionnaire; identify the target audience.
2. **Question Development:** Formulate questions clearly, understandably, and specifically; consider using various types of questions: open-ended, closed, rating scales, etc.
3. **Choosing the Survey Method:** Decide how the survey will be conducted (paper-based, online, by phone, in person, etc.); consider the characteristics of the target audience and the accessibility of the chosen method.
4. **Developing the Questionnaire Structure:** Define the order and logic of the questions; ensure a logical and consistent flow.
5. **Testing:** Conduct preliminary testing of the questionnaire with a small group of people to identify any issues with question wording or structure.
6. **Conducting the Survey:** Develop a data collection plan, including sampling, survey schedule, and participation motivation strategies; ensure confidentiality and anonymity of responses if necessary.
7. **Data Processing:** Determine data analysis methods according to the research goal; use appropriate statistical methods.
8. **Interpreting the Results:** Draw conclusions based on the data analysis; formulate recommendations or actions based on the results obtained.

As a result of the research, four questionnaires were developed: a university questionnaire, a digital skills questionnaire (student survey), an employer questionnaire, and a graduate questionnaire [6].

Rating assessment was made on the basis of analysis, processing and generalization of statistical indicators provided by the university according to the list of the rating model. The rating functions were determined on the basis of the approved criteria with the relevant weighting values.

For each criterion, indicators are formed to evaluate the activity of higher education institutions.

To measure the value of each parameter, indicators are formed, which allow assigning a certain number of points depending on the value of the parameter.

The data of the ranking report allow to assess the level of HEIs rating by the main criteria:

$$K = \sum_{i=1}^{4} K_i = K_1 + K_2 + K_3 + K_4 \tag{2}$$

where K_1 – is the criterion “Conditions for ensuring quality education”,
K_2 – criterion “Level of research activity”,
K_3 – criterion “Level of demand for graduates by employers”,
K_4 – criterion “Brand of the university”.

Accordingly, group G_1 – “Educational programs” of criterion K_1 – “Conditions for providing quality education” has a weight of 9.5 points and is calculated according to the following formula:

$$K_1 = \sum_{i=1}^{6} k_{1i} = 9{,}5 \tag{3}$$

where k_{1i} – is the coefficient of the i-th indicator.

Thus, the main steps of the algorithm for calculating HEI indicators for the formation of the national ranking of HEIs are defined:

Step 1. Input data for each HEI v for all indicators i.
Step 2: Calculation of the university's share indicator R_{vi} for each indicator i, where $v = \overline{1, N}, i = \overline{1, M}$.
Step 3: The maximum of R_{vi} values is determined for each indicator.

Step 3.1. $R_i = \max R_{vi} \Rightarrow v^* : R_{vimax} = 1$, where $v = \overline{1, N}, i = \overline{1, M}$.
Step 3.2. For the rest of the universities, the indicators R_{pvi} are calculated using the formula:

$$R_{pvi} = R_{vi}/R_{vimax}, \tag{4}$$

where R_{vimax} – is the maximum value of R_{vi} among all universities.

Step 4: Calculation of the overall ranking R_{vo} for each indicator by universities is calculated by the formula:

$$R_{vo} = R_{pi} \cdot k_i, \tag{5}$$

where k_i – Is the coefficient of the corresponding indicator.
Step 5: A ranked list of universities is compiled according to the following indicators:

- by indicators R_{vi}, where$v = \overline{1, N}, i = \overline{1, M}$. by groups

$$R_{vg} = \sum_{i=1}^{n} R_{vi}, \tag{6}$$

where $v = \overline{1, N}, i = \overline{1, M}$.
- by each criterion

$$R_{vkr} = \sum_{i=1}^{n} R_{vg} = \sum_{i=1}^{m} R_{vi}, \tag{7}$$

where $v = \overline{1, N}, i = \overline{1, M}$.

– for all criteria in the aggregate

$$R_v = R_{total} = \sum_{kr=1}^{4} R_{vkr} = \sum_{g=1}^{12} R_{vg} = \sum_{i=1}^{50} R_{vi} \tag{8}$$

where $v = \overline{1, N}$, $i = \overline{1, M}$.

Step 6: A ranked list of HEIs is compiled by the method of sorting in descending order by indicators:

$V = \{v1, v2, \ldots, vn\}$, where V – is a set of universities.

$R = \{R_{v1}, R_{v2}, \ldots, R_{vn}\}$, where R – is a vector of ranking scores, R_v – is the final score of university v.

Our goal is to find a permutation of indices σ such that:

$$R_{\sigma(1)} \geq R_{\sigma(2)} \geq \cdots \geq R_{\sigma(n)} \tag{9}$$

Step 7: Development of recommendations to HEIs that need to improve their indicators:

If $R_{vi} < \overline{R} \Rightarrow R_{vi} \uparrow$, where $\overline{R}$– is the average value of the indicator among all HEIs.

$$\overline{R} = \frac{1}{N} \sum\nolimits_{j=1}^{N} R_{vj}, \tag{10}$$

The logical-structural model of the automated university ranking system describes the sequence of operations and interaction of components—from data input to the generation of final results. Its goal is to ensure accuracy, efficiency, and transparency in evaluation.

A user (administrator, university representative, or analyst/commission member) inputs data through a web interface or uploads it from databases. The system then verifies the information for format compliance, data integrity, and duplicate entries. After that, indicators are weighted, data is normalized, and ranking calculation algorithms are applied.

As output, the system generates aggregated results—institutional rankings and an analysis of universities' strengths and weaknesses.

To ensure objective evaluation, the model covers the entire process—from data collection to final assessments and recommendations. It integrates data from different sources and includes four key modules: input data, processing, analytics, and results generation, ensuring a systematic and transparent workflow. The model for forming the integrated university ranking system is presented in Fig. 2 in the form of a Data Flow Diagram (DFD). Process "Formation of national ranking of universities" is responsible for the collection and processing of questionnaire data from different participants of the process and the formation of the final rating of universities. The system collects questionnaire data from universities, students, graduates, and employers, verifies its authenticity through commission representatives, processes the information according to predefined criteria, and generates a national university ranking, which is then provided to the relevant stakeholders.

The process of forming university rankings includes several sequential stages:

1. **Preparatory Stage** – the administrator creates a list of universities that will participate in the ranking.
2. **Input and Storage of Primary Data** – involves collecting information about university representatives and the universities themselves, commission members, and weight coefficients for various ranking criteria.
3. **Input and Storage of Survey Data** – collecting information based on surveys: university questionnaire, employer questionnaire, graduate questionnaire, and digital skills questionnaire.
4. **Processing of Experimental Data** – systematization and analysis of the collected information.
5. **University Ranking Formation** – calculation of indicators: values for specific indicators, values by criteria, values by groups, and the final overall score across all criteria.
6. **University Ranking** – generating the final list of universities based on the obtained scores and providing recommendations to specific institutions.

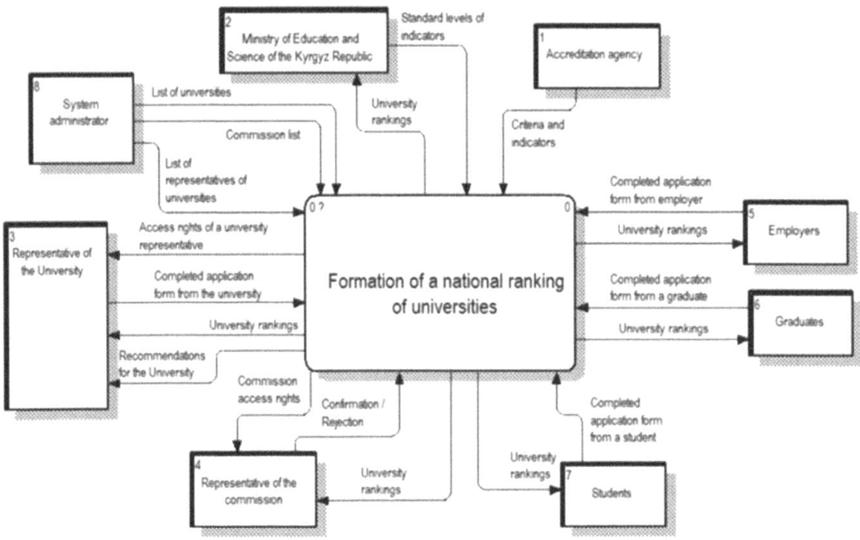

Fig. 2. Model for the Formation of the Integrated National University Ranking System of the Kyrgyz Republic in a Data Flow Diagram (DFD).

These stages are also presented as a decomposition of the DFD diagram – the model for the formation of the integrated national university ranking system of the Kyrgyz Republic (Fig. 3):

The formation of the national university ranking system of the Kyrgyz Republic involves the sequential implementation of a series of key processes that ensure the collection, verification, analysis, and systematization of data for the subsequent calculation of university rankings.

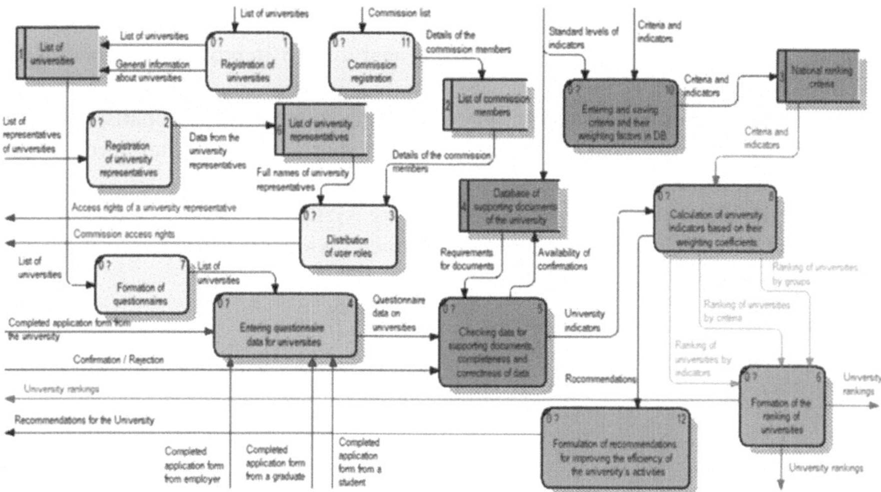

Fig. 3. Decomposition of the Formation of the Integrated University Ranking System of the Kyrgyz Republic in a Data Flow Diagram (DFD).

At the first stage, universities, their representatives, and commission members are registered. During this phase, basic information about universities and participants is collected, and lists of registered universities, commissions, and representatives are compiled for further interaction.

Next stage, user roles are assigned, granting each participant appropriate access rights to the system for completing surveys or verifying submitted data.

The subsequent stage involves the formation of questionnaires, designed to collect the necessary information from universities, students, graduates, and employers. The questionnaire templates are developed according to approved evaluation criteria.

After the questionnaires are prepared, the data input process takes place. At this stage, users complete the questionnaires, and supporting documents are uploaded into the system.

The submitted data then undergo mandatory verification, which includes validating the correctness of the questionnaires and the accompanying documents. As a result of the verification, the data are either confirmed or rejected, and an information set is prepared for the further calculation of university performance indicators.

Simultaneously, evaluation criteria and weighting coefficients are entered into the system database, ensuring the relevance of the ranking calculation methodology.

Based on the verified data and approved criteria, the calculation of university performance indicators is carried out, including intermediate assessments across levels, criteria, and indicators.

The formation of the final university ranking involves combining the calculated data into comprehensive rankings by levels, criteria, indicators, as well as compiling an overall ranking.

The final stage is the development of recommendations for universities, where the strengths and weaknesses of universities are analyzed to improve their operational effectiveness.

The university ranking formation process ensures an objective comparison of universities based on various parameters and forms the basis for the further development of educational institutions (Fig. 4).

The developed automated university ranking system (AURS) is based on the MVC (Model-View-Controller) architecture, which divides the system into three key components:

- Model – the application logic and data handling.
- View – displaying information to the user.
- Controller – processing user input and managing the model.

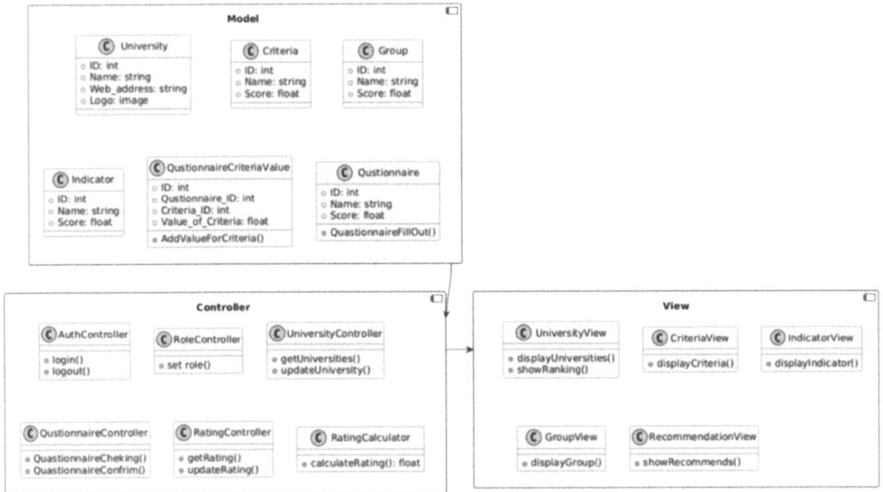

Fig. 4. MVC System Architecture

The AURS will create the most transparent ranking, which, in turn, will provide information about universities to prospective students, employers, and other interested parties. Figure 5 shows the main components of the system. The system consists of four key components: the user interface, data collection module, data processing module, reporting module, and database [7].

The results of the software implementation are presented in the form of a web application, which enables data collection through the generation of questionnaire forms, performs calculations based on specific criteria, verifies the authenticity of the data in a short time, and generates the ranking results.

The Automated University Ranking System (AURS) has undergone testing and experimental research at pilot universities: Zh.Balasagyn Kyrgyz National University (KNU), I.Razzakov Kyrgyz State Technical University (KSTU), Kyrgyz State University named after I.Arabayev (KSU), Kyrgyz National Agrarian University named after

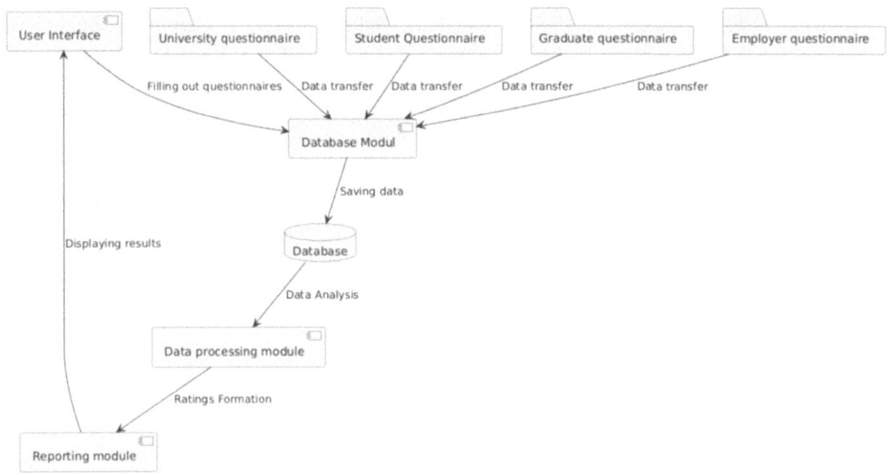

Fig. 5. AURS Components diagram

K.I.Skryabin (KNAU), and Osh Technological University named after M.M. Adyshev (OshTU). It has also been implemented at the independent accreditation agency IARC.

A ranked list of universities was formed based on criteria, groups, indicators, and the overall university rating. Overall university ranking results: I. Razzakov KSTU – 71.42, Zh.Balasagyn KNU – 57.47, KSU named after I.Arabayev – 55.61, KNAU named after K.Skryabin – 46.16 and OshTU named after M.Adyshev – 44.00.

4 Conclusion

The developed model of the national university ranking system of the Kyrgyz Republic allows:

- To increase the objectivity and transparency of university performance evaluations.
- To automate the data analysis and ranking formation process.
- To ensure the national system aligns with international standards.
- To form a foundation for the strategic development of universities and enhancing their competitiveness.

The automated ranking system will become an important tool for educational institutions, employers, and government bodies, contributing to the improvement of the higher education system in Kyrgyzstan.

References

1. Karazakova, Z., Sheranova, N.: The significance and role of evaluation through internal ranking of the activities of university teaching staff. Bull. Osh State Univ. **1**, 68–76 (2023)
2. Zhou, Y., Asipova, N.: Higher education system in the Kyrgyz Republic at the present stage. Sci. Herald Uzhhorod Univ. Ser. Phys. **55**, 2624–2633 (2024)

3. Chynybai, M.K., Arzybaev, A.M., Koshoeva, B.B., Bakalova, A.T.: Digital transformation of education using the example of KGTU. Izv. KGTU named after I. Razzakov **4**(52), 88–95 (2019)
4. Wang, G.: Construction of the index system for the integration of professional education and innovation entrepreneurship education in applied universities: based on the Kirkpatrick evaluation model. Int. J. Inf. Commun. Technol. Educ. **20**(1) (2024)
5. Akhatov, A.R., Mardonov, D.R., Nurmamatov, M.Q., Nazarov, F.M.: Improvement of mathematical models of the rating point system of employment. Research.SamDU.UZ (2021)
6. Koshoeva, B.B., Torobekov, B.T., Bakalova, A.T.: Design of business process algorithms for the automated system. High-Perform. Comput. Syst. Technol. **7**(1), 166–172 (2023)
7. Bakalova, A.T.: Development of an integrated automated national university ranking system of the Kyrgyz Republic. Ogaryov-Online 13(214) (2024)

Evaluating the Effectiveness of Voice Conversion Attacks on a ResNet-Based Text-Independent Speaker Verification System

Polina Karshieva(✉) and Ivan Rakhmanenko

Tomsk State University of Control Systems and Radioelectronics, 40 Lenina Avenue, 634050 Tomsk, Russia

polinakarshieva1@gmail.com, ria@fb.tusur.ru

Abstract. In the world of information technology, there is often the threat of fraud. One of the well-known tools of fraudsters is voice conversion. This paper models an attack on a text-independent speaker verification system using the following voice conversion methods: Hifi-VC, Diff-VC, DiffHier-VC and Vosk-VC. The research aims to evaluate the effectiveness of these voice conversion methods in spoofing attacks against a modern ResNet-based verification system. The English-language speech corpus TIMIT is used as input data for the voice conversion. It was used to obtain four new datasets, which were used in modeling an attack on the text-independent speaker verification system. The entire experiment is divided into three main phases, which are distinguished by the ratio of original audio recordings to fake ones. The value of Equal Error Rate (EER) was used as the criterion for evaluating effectiveness when testing the text-independent speaker verification system. According to the results of the modeling, it was concluded that the conversion methods used are effective in attacking the text-independent speaker verification system because the EER value increased with the increasing number of fake audio recordings. The DiffHier-VC method was the most efficacious, elevating the EER by 325% and 525% in Stages 2 and 3 of the experiment, respectively. The statistical significance of these results was confirmed by the Wilcoxon signed-rank test. The results of the study can be used in further training of speaker verification systems.

Keywords: Speaker Verification · Voice Conversion · Attacks On The Speaker Verification System · Equal Error Rate · Speech Processing · Voice Spoofing

1 Introduction

With the growth of online services and remote technologies, security and safety of user identity verification are becoming critical aspects of modern society. Nevertheless, as information technology develops, new opportunities for fraudsters also appear. Many banking organizations use speaker verification [1]. Human voice is an excellent biometric parameter that provides a high degree of individual uniqueness. However, speaker verification has a number of disadvantages [2]. The biggest one is the extreme vulnerability to spoofing [3, 4].

V. Jordan et al. (Eds.): HPCST 2024, CCIS 2919, pp. 175–187, 2026.
https://doi.org/10.1007/978-3-032-20325-0_14

One of the well-known ways to fraud a speaker verification system is voice conversion [5, 6]. At the current moment, there are numerous methods for converting voice characteristics of a speaker. This problem requires continuous improvement of protection mechanisms. Most practical applications of voice conversion models are used purely for entertainment purposes. However, the software implementation of such methods is openly accessible to any Internet user. In such a scenario, basic programming skills are enough for fraudulent purposes. By using voice spoofing, fraudsters can gain access to accounts and credit cards.

The goal of this research is to study existing voice conversion methods and their effectiveness in attacks on the modern speaker verification system. Understanding the specifics of such methods and their influence on security may allow us to identify potential vulnerabilities in verification systems. The results of the study can be used in further design of verification systems and their training in particular.

Achieving this objective requires the following steps:

- Select voice conversion methods suitable for the aims and objectives of the research.
- Select a dataset for further conversion.
- Generate new datasets using the selected voice conversion methods.
- Conduct a modelling attack on the text-independent speaker verification system.
- Evaluate the effectiveness of the selected methods.

In Sect. 2, the design of the experiment and the voice conversion process are described. Further, Sect. 3 provides the modeling results of the attack on the speaker voice verification system, statistically validated. The modeling results are analyzed in Sect. 4. It also evaluates the effectiveness of voice conversion attacks. Finally, Sect. 5 presents recommendations for practical application of the research results.

2 Experimental Design

2.1 Choice of Voice Conversion Methods

It was necessary to select several voice conversion (VC) methods further utilized to model the attack on the text-independent speaker verification system. A total of four methods were selected. These are Hifi-VC [7], Diff-VC [8], DiffHier-VC [9], and Vosk-VC [10]. Key criteria utilized in the selection of the methods were as follows: Any-to-Any conversion capability and novelty of the methods.

Hifi-VC [11] adapts the HiFi-GAN architecture to directly generate waveforms from intermediate features, bypassing the explicit mel-spectrogram creation stage. Unlike previous approaches that used HiFi-GAN solely as a vocoder, this method integrates it into a single module that performs both decoding and synthesis, which is its key innovation.

Diff-VC [12] is a diffusion-based voice conversion method that implements its decoder within a Diffusion Probabilistic Model (DPM) framework. While this delivers high quality, it inherently suffers from slow inference due to the iterative sampling process. To address this, the authors developed a fast sampling scheme that significantly reduces the number of iterations required without compromising output quality.

DiffHier-VC [13] introduces a hierarchical system based on two diffusion models. The first, DiffPitch, generates an F0 contour conditioned on the target voice, while the second, DiffVoice, transforms the speech using this F0 and a converted mel-spectrogram as a prior. This hierarchical approach, aided by a source-filter encoder, effectively disentangles speech parameters to enhance conversion quality.

Vosk-VC [14] is a modified version of the VITS-based Quick-VC [16]. Key improvements include the use of TPRLS GAN to better preserve temporal properties in synthesized speech and the replacement of Hubert with Contentvec. These modifications enable the model to achieve superior results compared to the original Quick-VC.

These methods were published after 2020. For the goal of this research, modern methods are favored because they tend to incorporate the latest scientific and technological advances. Such tendencies lead to improved accuracy and performance compared to outdated methods.

The Any-to-Any conversion capability directly affects the independence of the experiment. The relevant condition states that test audio recordings are not present in the training dataset. Essentially, this simulates a scenario where an attacker has recorded the target's voice, but the available data is insufficient to fully train a voice conversion model for that specific speaker.

2.2 Experimental Stages

For the verification system subjected to spoofing attacks, a ResNet-based text-independent speaker verification system [12] was used. This system has been pre-trained on the VoxCeleb dataset [13], which consists of approximately 7000 different speakers. This system solves the verification problem in an open speaker set environment, where the test data does not include speakers from the training set.

To evaluate the effectiveness of the voice conversion attacks, the Equal Error Rate (EER) was chosen as the metric for testing the verification system. EER represents the point where the likelihood of a false acceptance (impostor passing) equals the likelihood of a false rejection (legal user being denied) [1].

The experimental results are recorded during random attacks on the text-independent speaker verification system. If EER increases, the corresponding voice conversion method is considered effective. The following stages of the experiment were emphasized:

1. Stage 1: only audio recordings from the original dataset are used. This stage involves one experiment file containing 3360 tests. Each test represents a speaker verification procedure for two audio recordings. The number of the tests verifying a legal speaker (two audio recordings of the same speaker) and the number of tests verifying two different speakers are distributed in a 1:1 ratio.
2. Stage 2: this stage uses audio recordings from both the original dataset and the converted dataset. Four experiment files are generated, one file for each conversion method. Each experiment file should maintain a 1:1 ratio of audio recordings from the original dataset and audio recordings from the converted dataset. At this stage, attacks are introduced, representing the verification of an original audio recording of a speaker against a converted one (of the same speaker). It is important to notice that spoken phrases for the original and converted audio recordings are different. The total number of verifications is similar to Stage 1.

3. Stage 3: this stage combines the tests from Stage 1 with an equal number of attacks using fake audio recordings. This results in four experiment files, one file for each conversion method. The total number of tests in each experiment file should be 6720 exactly.

The experiment file is a text document that contains multiple lines. Each line represents a separate pair of audio recordings (for example, a random recording of speaker №1 and a random recording of speaker №2). Each line has a label at the beginning: label «1» denotes a legitimate pair of audio recordings, and label «0» denotes an attack. A legal pair consists of two audio recordings from the same speaker, while an attack involves either a comparison of recordings from different speakers or a comparison of an original audio recording with a converted (fake) recording.

The result will be a total of nine experiment files for three stages and four different voice conversion methods.

2.3 Voice Conversion

As described in Sect. 2.2, modeling an attack on the verification system requires creating experiment files that include converted audio recordings. A total of nine files need to be generated. To achieve this, new datasets were created using the selected conversion methods [7–10].

To model the attacks, trained conversion models were applied to audio recordings from the TIMIT dataset [14]. A total of 168 male and female speakers were selected for voice spoofing attacks. Each speaker has 10 audio recordings with different phrases spoken.

The source-target audio pairs were processed as follows: the source is an audio recording from which the linguistic content (i.e., words) is extracted, and the target is an audio recording of the speaker whose voice is to be replicated. For each target speaker, every audio recording was used as the target, while a random audio recording from a different speaker served as the conversion source.

After performing the conversions, four new datasets were generated, each containing 1680 audio files, one dataset for each conversion method. In total, 6720 converted audio recordings were obtained.

3 Results

3.1 Attack Modeling

As defined in Sect. 2.2, the EER, obtained from testing the verification system [12], was used to evaluate the effectiveness of the selected methods. An experiment was designed to determine the effectiveness of each method. This experiment studies how EER changes after using converted audio recordings in testing the verification system.

In Stage 1, where no converted voice recordings were used, an EER value of 0.004 was obtained (see Fig. 1).

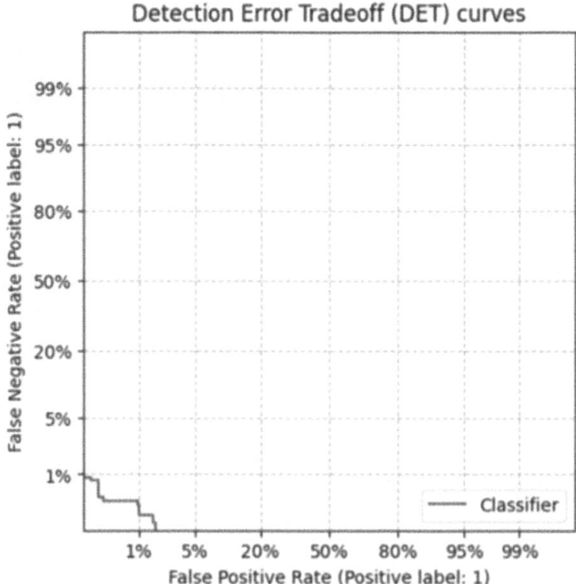

Fig. 1. DET-curve of the speaker verification system for Stage 1 of the experiment

Stage 2 involved evaluating the effectiveness of attacks on the verification system using the Hifi-VC (see Fig. 2), Diff-VC (see Fig. 3), DiffHier-VC (see Fig. 4), and Vosk-VC (see Fig. 5) methods. For the DiffHier-VC and Vosk-VC methods, the EER values increased by 325% and 25%, respectively, compared to Stage 3.

In Stage 3, the EER values increased for the Hifi-VC (see Fig. 6), Diff-VC (see Fig. 7), DiffHier-VC (see Fig. 8), and Vosk-VC (see Fig. 9). Respectively, the values were changed by 25%, 50%, 525%, and 75% compared to Stage 1, and by 25%, 50%, 50%, and 40% compared to Stage 2.

The results of all stages of the experiment for each method are presented in Table 1. The numerical values in the table are the EER values.

3.2 Statistical Significance of Results

Next, we needed to verify that the obtained results were statistically significant and that the chosen voice conversion methods genuinely affected the text-independent speaker verification system's vulnerability.

To achieve this, the Wilcoxon signed-rank test was used. This test is suitable for comparing two dependent samples that do not necessarily follow a normal distribution.

As shown in Sect. 3.1, the third stage of the experiment demonstrated better performance for all methods, as the number of attacks was higher compared to the second stage. This resulted in more frequent classification errors by the verification system. Therefore, this stage was selected to test the significance of the results.

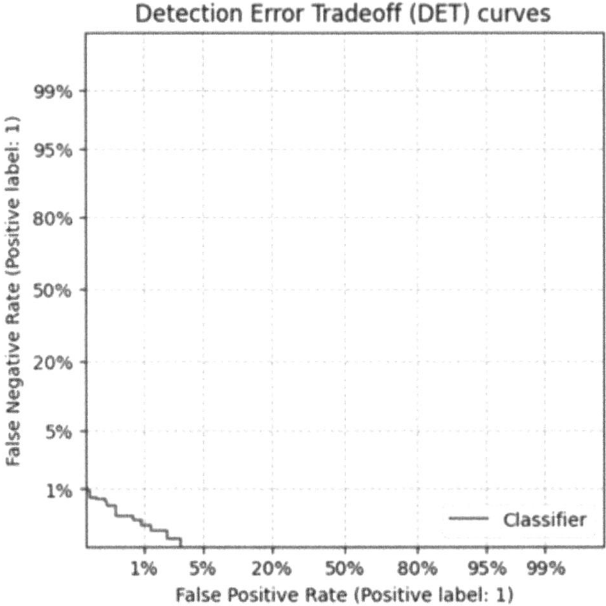

Fig. 2. DET curve of the speaker verification system for Stage 2, using Hifi-VC

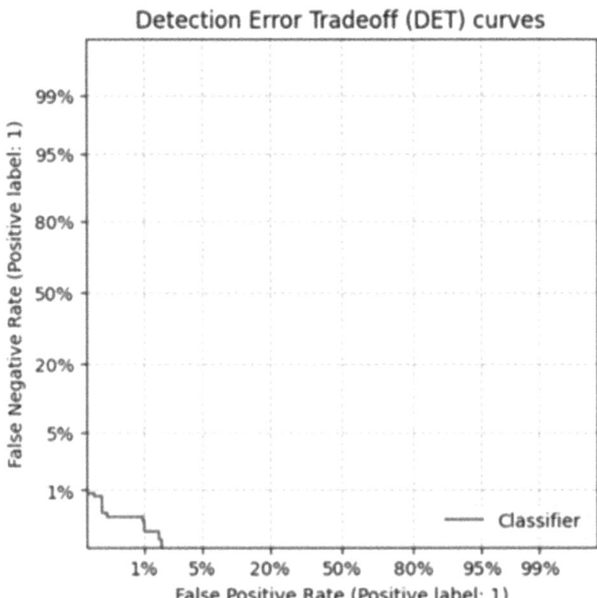

Fig. 3. DET curve of the speaker verification system for Stage 2, using Diff-VC

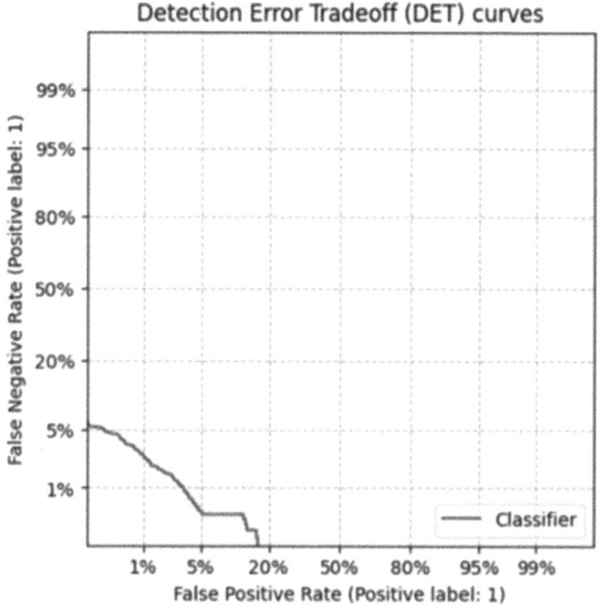

Fig. 4. DET curve of the speaker verification system for Stage 2, using DiffHier-VC

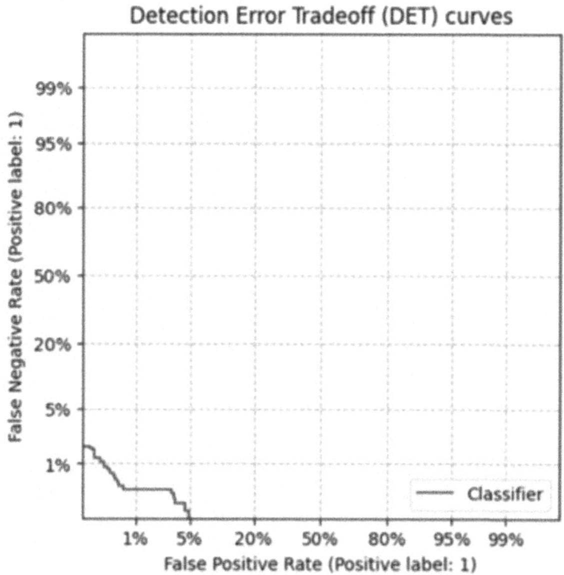

Fig. 5. DET curve of the speaker verification system for Stage 2, using Vosk-VC

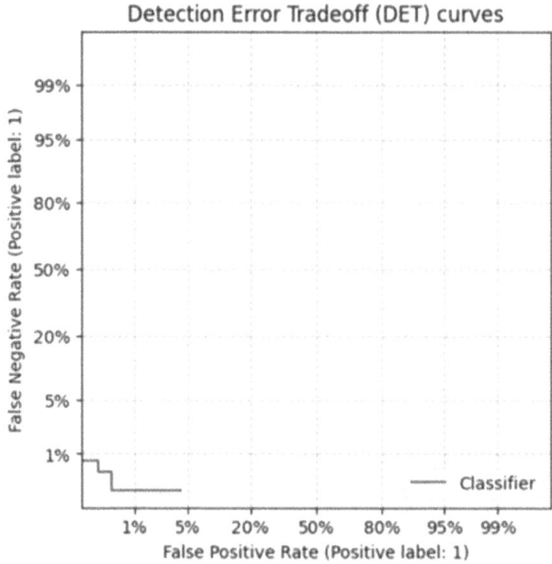

Fig. 6. DET curve of the speaker verification system for Stage 3, using Hifi-VC

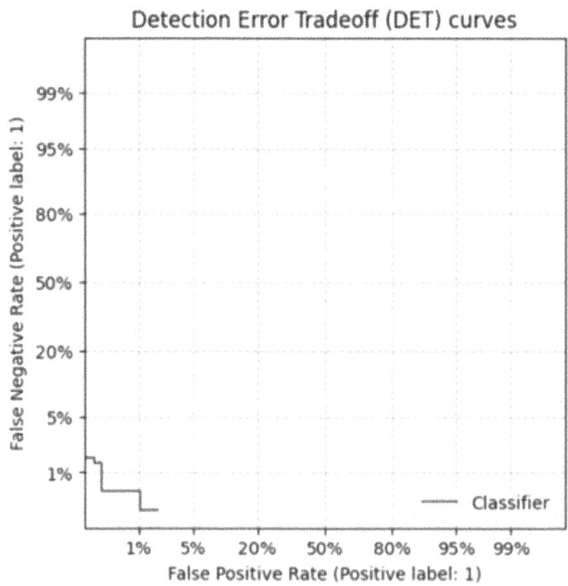

Fig. 7. DET curve of the speaker verification system for Stage 3, using Diff-VC

Adhering to the procedure of the third stage, 9 further experimental iterations were performed. Each iteration was based on a new set of random audio recording pairs. The outcomes are detailed in Table 2.

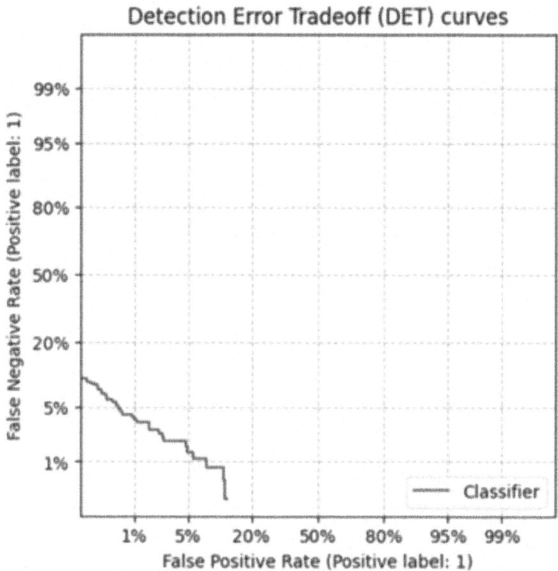

Fig. 8. DET curve of the speaker verification system for Stage 3, using DiffHier-VC

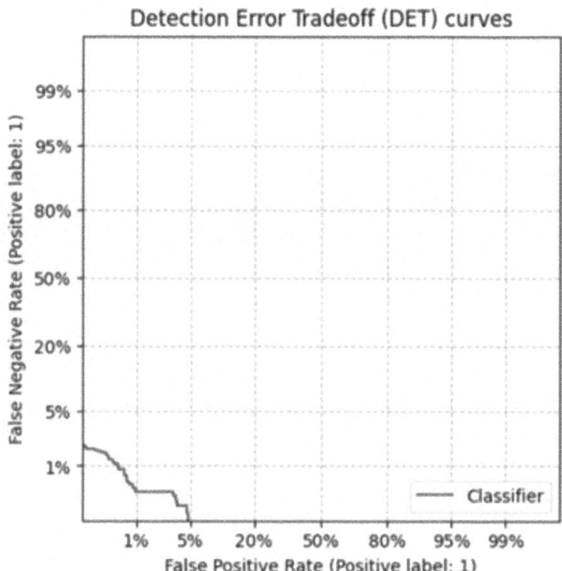

Fig. 9. DET curve of the speaker verification system for Stage 3, using Vosk-VC

The hypotheses tested were:

Table 1. EER for voice conversion attacks on ResNet-based text-independent speaker verification system

Voice conversion method	Stage 1	Stage 2	Stage 3
Hifi-VC	0.004	0.004	0.005
Diff-VC		0.004	0.006
DiffHier-VC		0.017	0.025
Vosk-VC		0.005	0.007

- The EER values obtained using the converted audio recordings (stage III) are less than or equal to those obtained in stage I of the experiment.
- The EER values obtained in stage III are greater than those obtained in stage I.

The EER values obtained from the first stage, prior to the introduction of converted audio recordings, served as the baseline for comparison. The significance level (α) was set at 0.05, representing an acceptable Type I error rate.

The empirical value of the Wilcoxon signed-rank test is calculated using the formula:

$$W_{stat} = min(T_{+}, T_{-}), \quad (1)$$

where W_{stat} represents the test statistic, calculated as the minimum value between T_{+} and T_{-}, T_{+} denotes the sum of the ranks of the positive differences between paired observations, T_{-} denotes the sum of the ranks of the negative differences between paired observations.

Table 3 presents the critical and empirical values of the Wilcoxon signed-rank test for the corresponding sample comparisons.

Based on the obtained values of the signed-rank test, since $W_{stat} < W_{crit}$, the results are statistically significant. Therefore, we reject the null hypothesis, concluding that the EER values obtained using the converted audio recordings (stage III) are statistically significantly greater than those obtained in stage I of the experiment.

The statistically significant results allow us to conclude that the outcomes presented in Sect. 3.1 are not random.

Table 2. EER values

Iteration	Hifi-VC	Diff-VC	DiffHier-VC	Vosk-VC
1	0.005	0.006	0.025	0.007
2	0.004	0.006	0.027	0.006
3	0.005	0.006	0.026	0.009
4	0.006	0.007	0.025	0.007
5	0.005	0.006	0.026	0.007
6	0.005	0.005	0.025	0.007
7	0.006	0.006	0.024	0.007
8	0.006	0.008	0.028	0.009
9	0.005	0.006	0.025	0.008
10	0.006	0.007	0.024	0.007

Table 3. Wilcoxon signed-rank test values for the experimental results

Criterion	Hifi-VC	Diff-VC	DiffHier-VC	Vosk-VC
W_{crit}	8	10	10	10
W_{stat}	0	0	0	0

4 Discussion

In this paper, modern and close-to-real developments and research were used. The low EER value that was obtained in Stage 1 demonstrates the accuracy of the verification system [11]. Thus, the results obtained are considered reliable.

In Stage 2, both DiffHier-VC [9] and Vosk-VC [10] showed positive results, with EER values increasing by 325% and 25%, respectively. DiffHier-VC proved to be the most effective method, exhibiting an EER value 13 times higher than that of Vosk-VC. In Stage 3, all methods proved effective, with EER values rising by up to 75% compared to Stage 1. However, DiffHier-VC again showed superior results, achieving a 525% increase in EER. The statistical significance of these results was confirmed using the Wilcoxon signed-rank test.

These results demonstrated that the voice conversion methods are effective at attacking the verification system, as the EER increased with the number of fake recordings, indicating a rise in verification errors.

5 Conclusion

This paper evaluates the effectiveness of voice conversion attacks on speaker verification system. As a result of the research, it was found that the selected voice conversion methods are effective, as evidenced by a rising EER correlated with an increasing number of fake audio recordings. The most successful method was DiffHier-VC.

In the process, 4 new datasets were created. These datasets consist of converted audio recordings that were effectively used in the attacks. It is planned to use the created datasets in future studies. The research will be aimed at direct improvement of verification systems and their resistance to fake audio recordings.

Acknowledgments. This work was performed as part of the TUSUR Development Program for 2025–2036, which falls under the Strategic Academic Leadership Program "Prioritet 2030" (Priority 2030).

References

1. Rakhmanenko, I., Shelupanov, A., Kostyuchenko, E.: Automatic text-independent speaker verification using convolutional deep belief network. Comput. Opt. **44**(4), 596–605 (2020)
2. Lindberg, J., Blomberg, M.: Vulnerability in speaker verification a study of technical impostor techniques. In: Proceedings of the 6th European Conference on Speech Communication and Technology (Eurospeech 1999), pp. 1963–1966. ISCA, Budapest (1999)
3. Evsyukov, M., Putyato, M., Makaryan, A., Nemchinova, V.: Protection methods in modern voice authentication systems. Caspian J. Control High Technol. **3**(59), 84–92 (2022)
4. Faundez-Zanuy, M., Hagmuller, M., Kubin, G.: Speaker verification security improvement by means of speech watermarking. Speech Commun. **12**(48), 1608–1619 (2006)
5. Toda, T., Black, A., Tokuda, K.: Voice conversion based on maximum-likelihood estimation of spectral parameter trajectory. IEEE Trans. Audio Speech Lang. Process. **8**(15), 2222–2235 (2007)
6. Wu, C., Hsia, C., Liu, T., Wang, J.: Voice conversion using duration-embedded bi-HMMs for expressive speech synthesis. IEEE Trans. Audio Speech Lang. Process. **4**(14), 1109–1116 (2006)
7. Kashkin, A., Karpukhin, I., Shishkin, S.: HiFi-VC: high quality ASR-based voice con version. In: Proceedings of the 12th ISCA Speech Synthesis Workshop, pp. 100–105. ISCA, Grenoble (2023)
8. Popov, V., Vovk, I., Gogoryan, V., Sadekova, T., Kudinov, M., Wei, J.: Diffusion-based voice conversion with fast maximum likelihood sampling scheme. Preprint at https://arxiv.org/abs/2109.13821
9. Choi, H., Lee, S., Lee, S.: Diff-HierVC: diffusion-based hierarchical voice conversion with robust pitch generation and masked prior for zero-shot speaker adaptation. In: Proceedings of the 24th Annual Conference of the International Speech Communication Association, pp. 2283–2287. ISCA, Dublin (2023)
10. Vosk-VC: basic zero-shot voice conversion model. https://github.com/alphacep/vosktts/tree/master/vc, last accessed 2024/06/24
11. Guo, H., Liu, C., Ishi, C.T., Ishiguro, H.: QuickVC: any-to-many voice conversion using inverse short-time fourier transform for faster conversion. In: Automatic Speech Recognition and Understanding Workshop, pp. 1–7. Taipei (2023). https://doi.org/10.48550/arXiv.2302.08296

12. Chung, J.S., et al.: In defence of metric learning for speaker recognition. Preprint at https://arxiv.org/abs/2003.11982
13. VoxCeleb Acoustic-Phonetic Continuous Speech Corpus. https://www.robots.ox.ac.uk/~vgg/data/voxceleb. Accessed 29 Jun 2024
14. TIMIT Acoustic-Phonetic Continuous Speech Corpus. https://paperswithcode.com/dataset/timit. Accessed 20 Jun 2024

Information and Computing Technologies in Automation and Control Science

Collector Module of Mechatronic Microclimate Control System

Alexey Dolmatov(✉), Maria Dolmatova, Evgeniy Godovnikov, and Egor Safonov

Ugra State University, Chekhova Street, 16, Khanty-Mansiysk 628011, Russia
adolmatov@bk.ru

Abstract. The work is devoted to the creation of an intelligent device for individual watering of house plants, functioning in a distributed microclimate control system. The structure of the mechatronic system is presented and the levels of its deployment are designated. An additional motivation for building a distributed system was the study of synchronization of the operation of mechatronic modules through network communications. The microclimate system is represented by a set of real and virtual modules, each of which inherits the properties of a network intelligent device. The interaction of information and executive modules in a distributed system is carried out via a wireless network interface. Virtual modules of the system can be located both on computers in the local network and in cloud services on the Internet. CAD technologies were used in the design of the individual plant irrigation collector module, and the production of its structural elements was carried out on the basis of PLA plastic prototyping. Distribution of nutrient liquid to plants is carried out by electromagnetic valves and a digital flow meter controlled by an ESP8266 microchip. The configuration of the operating modes of the collector module is based on a Web service, the server of which is deployed on the built-in microcontroller.

Keywords: Network System · Microclimate Control · Irrigation · Synchronization · Embedded Computing Platform · 3D Prototyping

1 Introduction

The development of microprocessors and network technologies has led to the formation of a new class of measurement and control instruments [1, 2]. Embedded computing platforms have greatly expanded the intelligence of devices and made it possible to coordinate their operation through short network messages [3, 4]. To fully realize the potential of intelligent modules in distributed control systems for fast-moving processes, it is important to develop new techniques and methods for reducing the time required to make decisions and transmit control actions in a communication environment [5, 6].

Modern homes have developed local networks and access to the global Internet. The communication environment necessary for the operation of "smart" distributed control systems has been created here. On the one hand, it can be used to implement the

V. Jordan et al. (Eds.): HPCST 2024, CCIS 2919, pp. 191–201, 2026.
https://doi.org/10.1007/978-3-032-20325-0_15

"Smart Home" concept, and on the other hand, to build a laboratory testing ground for studying the interaction of network intelligent modules and improving the protocol for the operation of a distributed control system in real time, taking into account the time constant characteristic of fast-moving processes [7–9].

Currently, there are several families of embedded microcontrollers with a developed ecosystem of software support. They have different computing and functional capabilities, which allows optimizing the cost of hardware in the process of designing mechatronic modules. The ESP8266 family of microcontrollers is primarily used to organize wireless network communications and allows to minimize development costs. At the same time, the significant computing power of these microchips can be used to embed a Web service into a mechatronic module and implement a synchronous software control architecture.

The purpose of the work is to plan the structure of a network system for monitoring the microclimate of houseplants and to design a collector module for their individual irrigation.

2 Network Microclimate Control System

Irrigation of plants is aimed at improving the supply of moisture and nutrients to the roots, reducing the temperature of the surface layer of air and increasing its humidity. In the case of house plants, it is necessary to include the individual habitat of each specimen, as well as the characteristics of its moisture, nutrition, lighting and temperature range of growth. Therefore, the tasks of the developed system for monitoring the microclimate of house plants can be defined as follows:

- individual irrigation;
- individual control of soil and air humidity, air temperature;
- local control of illumination of several plants;
- automation of the construction of a connection of the type "plant condition - microclimate parameters";
- forecast of quantitative parameters of irrigation and lighting for correction of plant microclimate;
- adaptation of the microclimate control system to seasonal changes in the external environment and the evolution of the hardware functions of the sensors;
- monitoring of microclimatic parameters of individual plants and remote control of the system.

Indoors, plants are usually placed near windows, and the water supply for the irrigation system can be either centralized or autonomous. The peculiarities of irrigation of house plants include low intensity of water supply to maintain a given level of soil moisture. Therefore, the watering process can be carried out in several stages with alternating supply of water to individual plants in small doses throughout the day. Dividing plants into window groups allows the use of local lighting control devices for simultaneous additional illumination of several plants. Figure 1 shows a structural diagram of a plant microclimate control system, constructed taking into account the listed features.

The source selection modules in the climate control system determine the type of liquid (water or feed solution) and the method of its delivery. In the case of a centralized supply, the water pressure is determined by an external source, and when feeding liquid from an autonomous tank, the source selection module uses its own pump to create pressure in the main line. Collector modules are used to distribute liquid between individual plants.

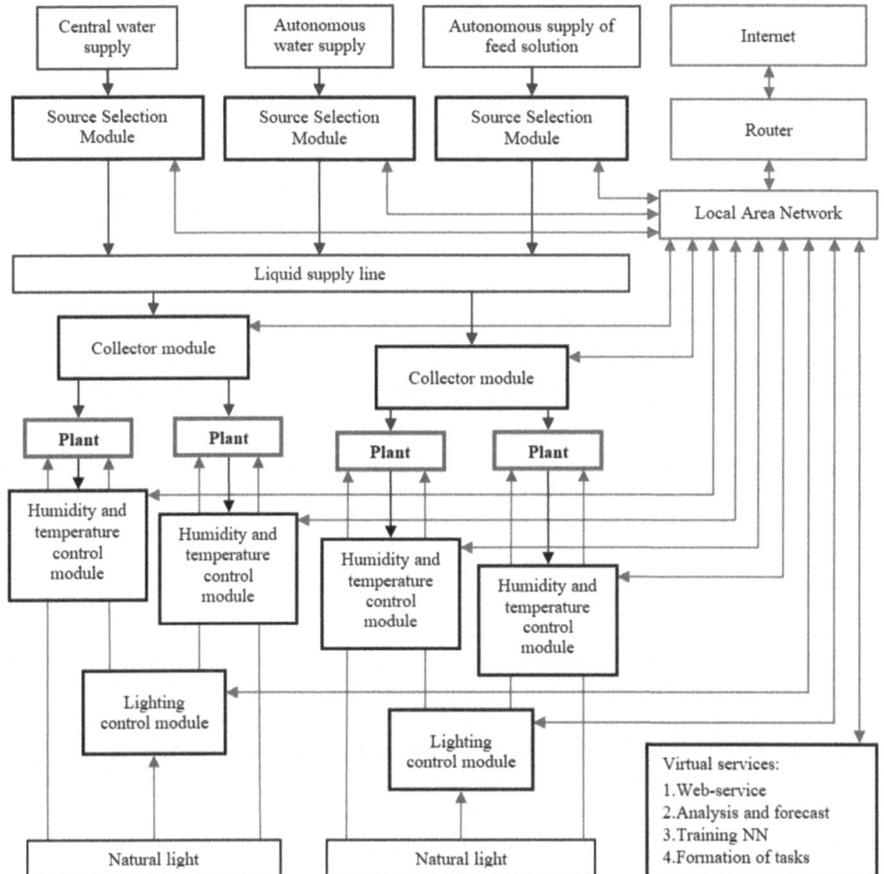

Fig. 1. Structure of a mechatronic system for controlling the microclimate of house plants.

They are equipped with a liquid flow sensor, which is used to control the dose during sequential irrigation of plants. Individual humidity and temperature control modules regularly collect data on soil and surface air humidity and measure the air temperature around the plant. The purpose of the illumination control module is to determine the dose of natural light during the day and to organize additional lighting at night.

In a distributed indoor plant microclimate control system, each module inherits the basic properties of networked intelligent devices, which can be designated as follows:

- is based on an embedded computing platform, the hardware of which is optimized for the specific methods of processing the recorded data;

- the exchange of data or control signals with neighbors in the control system is carried out only through the network interface;
- is ready to initiate and enter into neighborly relations with other mechatronic modules to build distributed control systems;
- is capable of interacting with its neighbors in the control system via unique and group network messages, and is also ready to participate in the operation of an unlimited number of similar systems throughout its entire service life;
- has its own storage of data specific to it and supports during operation a mechanism for their identification in agreement with its neighbors;
- supports the mechanism of synchronization with the world time system with a given accuracy;
- strives to minimize your own network traffic during the data registration process;
- provides permanent authorized access to its own data and methods of processing it for current and former neighbors;
- supports digital twin technology and organizes a Web service for publishing information about the intended purpose and providing interaction tools.

The specified protocol allows any of the mechatronic modules to initiate the construction of a distributed system of sensors and actuators, and present it in the form of a logical diagram. The scheme is created for a limited period of time, during which participants do not have to agree to be included in other systems. The responsibility for the physical organization of modules in a network system lies with its user. During the logic diagram construction process, the initiating module uses a neighbor establishment procedure, by which it communicates to other members the system GUID and the group broadcast IP address, and assigns to each an action program tied to sets of incoming and outgoing messages. Incoming messages are used by modules to synchronize the steps of their individual program, and outgoing messages initiate actions of neighbors. During operation, each mechatronic module stores data specific to it in its own database. However, the system GUID and the timestamps of its own real-time clock are used to identify records. Synchronization of the clock with world time is performed using the NTP or PTP protocol [10]. The choice of protocol depends on the requirements for clock accuracy. Identification of records in module databases enables an external entity to reconstruct the logical structure of the system's operation at any time and combine information for its own analysis and processing. During operation, the mechatronic module can be used in different systems. The module collects data only within the framework of the logic scheme that is currently in effect. However, analysis of previously collected information can be carried out at any time. Therefore, the module has the ability to simultaneously exchange data with current and former neighbors. In order to optimize such information exchange, the mechatronic module must have digital twins in the global network, supporting the virtualization protocol and data synchronization with backup servers. To support the API, the mechatronic module must provide client parts of the network service as object model classes in popular programming languages, publishing these tools along with documentation on a cloud Web-server.

The deployment of a microclimate control system for houseplants involves the following stages:

- organization of work of collector modules with an open control system. At this stage, the microclimate control system can function on the basis of a single collector module, the Web interface of which facilitates programming of the daily watering regime of plants. The watering regime is determined by the volume and number of doses of water or an aqueous solution. The watering history of each plant is saved in the non-volatile memory of the built-in controller and is displayed on the device's website.
- organization of a closed control system based on individual climate sensors of plants and group lighting control devices. At this stage, a given level of soil moisture is used to control irrigation, and additional illumination is carried out based on the deviation of the daylight level from the norm. The temperature and humidity of the air around the plant are used as disturbance factors of the control system;
- organization of an adaptive control system based on the connection type "plant condition - its microclimate parameters". At this stage, a plant condition monitoring service is deployed in the microclimate control system, using regular Web-survey of the user, and forecasting of irrigation and lighting doses is applied using artificial intelligence.

Some functions of the microclimate control system do not require the use of special hardware and can be implemented as virtual services that rely on data from the measuring modules and set the operating parameters of the system's actuators. Such services include the general Web interface of the system, training services and the operation of artificial neural networks used to predict optimal soil moisture and illumination of individual plants. Virtual modules of the system can be located both on general-purpose computers in a local network and be transferred to cloud services on the Internet.

3 Hardware for Collector Module

The initial stage microclimate control system is based on a collector module for individual plant irrigation. Its design uses pipeline fittings used in reverse osmosis membrane water purification devices (Fig. 2). Borrowing elements of advanced water purification technology has made it possible to minimize financial and time costs for developing a collector module for a microclimate control system for houseplants. The control of liquid flow in the manifold module is based on electromagnetic valves with collet quick-release connections for a plastic tube with a diameter of 6.5 mm. The organization of the manifold pipeline involves various types of collet fittings with the same diameter as of the connecting tube.

Such pipeline fittings are easily installed inside modern baseboards, which makes it easy to deploy a stationary irrigation system for house plants without significant repairs.

For compact placement of valves inside the manifold module, special equipment has been developed that combines them into a group of four devices (Fig. 3). The manifold module uses two solenoid valve assemblies to control the flow of liquid to eight plants.

The design of the tooling and other structural elements of the collector module was carried out using CAD technologies (Fig. 4-a), and 3D prototyping based on PLA plastic was employed for their production (Fig. 4-b).

A plastic tube with a diameter of 6.5 or 10 mm, designed for a maximum pressure of 10 bar, can be used to supply liquid to the collector module. Liquid drainage to plants is facilitated by a transparent silicone tube with a working pressure of up to 2 atm.

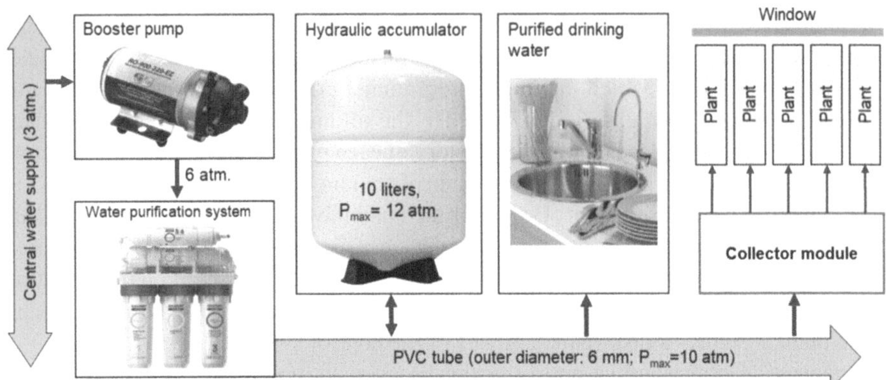

Fig. 2. Combined operation of drinking water filtration and plant irrigation.

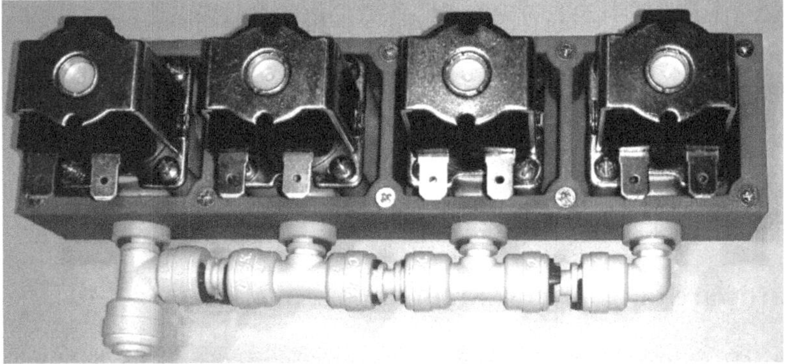

Fig. 3. Grouping of manifold module valves using tooling.

The flow of liquid through the collector module is monitored by a digital flow meter with a resolution of 450 pulses per liter. The arrangement of pipeline elements inside the collector module is shown in Fig. 5. The electronic components of the collector module are separated from the pipeline by a special partition and a cover with a hermetic seal (Fig. 6).

The composition and interaction diagram of the electronic components of the collector module are shown in Fig. 7. The operating voltage (12 V) and current (0.4 A) of the electromagnetic valves are provided by an external power supply unit. Current control in the circuit of each valve is carried out using an assembly of power keys (N-FET – Zelo-module) built on shift register microcircuits and an SPI interface. The Wi-Fi Slot board based on the ESP8266 microcircuit with a clock frequency of 80 MHz and a Flash memory capacity of 2 MB is used as an embedded microcontroller platform of the collector module.

ESP8266 supports the wireless Wi-Fi 802.11 b/g 2.4 GHz interface in the following modes: client (STA); access point (AP); client + access point (STA + AP), has a hardware implementation of the SPI interface, a nominal operating voltage of 3.3 V at a maximum

a)

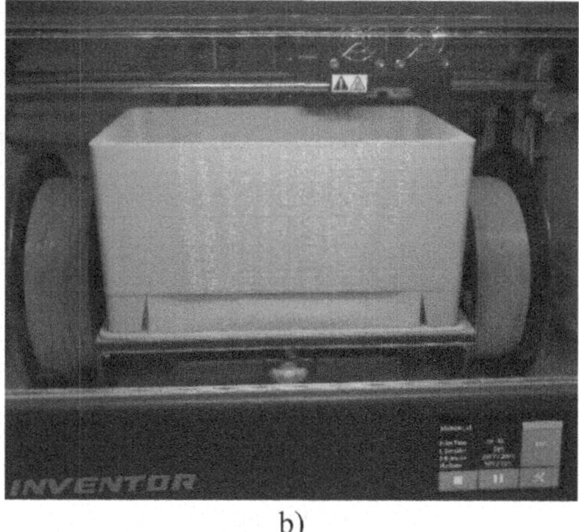

b)

Fig. 4. Design (a) and production (b) of structural elements of a collector module.

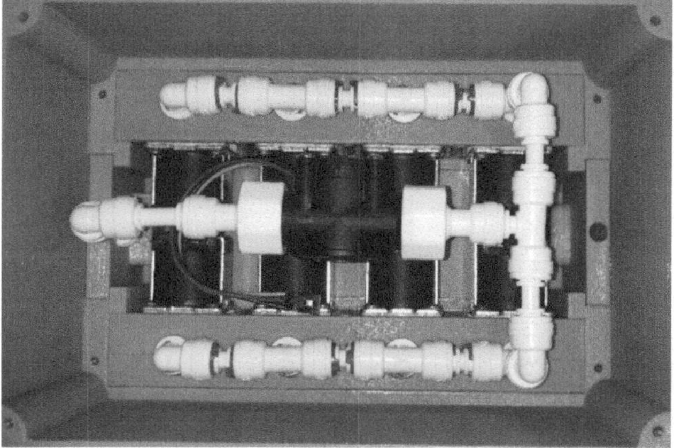

Fig. 5. Installation of the pipeline inside the collector module.

current of 1 A. The dimensions of the microcontroller platform are 51 × 51 mm. To supply power to the platform, a mini-board of a linear voltage regulator (dimensions 25 × 25 mm) is used, which reduces the input 12 V to a level of 3.3 V.

In addition to the external source, the microcontroller is powered by a built-in battery with a 5-volt charger. The electronic components of the collector module include an SD-card reader, which increases its non-volatile memory to 32 GB and guarantees autonomous storage of data reflecting the implementation of the plant irrigation program. In general, the hardware resources of the microcontroller platform are sufficient to support the basic properties of a network intelligent device.

The ESP8266 software is developed in the Arduino IDE environment and, using libraries, implements an object-oriented approach to controlling MOSFET keys via the SPI interface, synchronizes the mechatronic module clock via the NTP protocol, provides

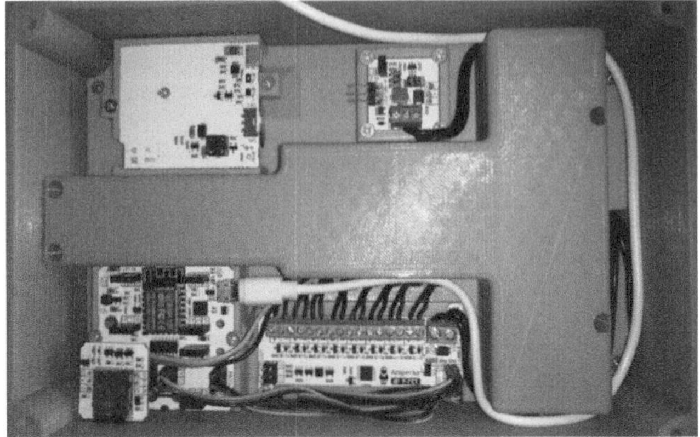

Fig. 6. Electronic component level of the collector module.

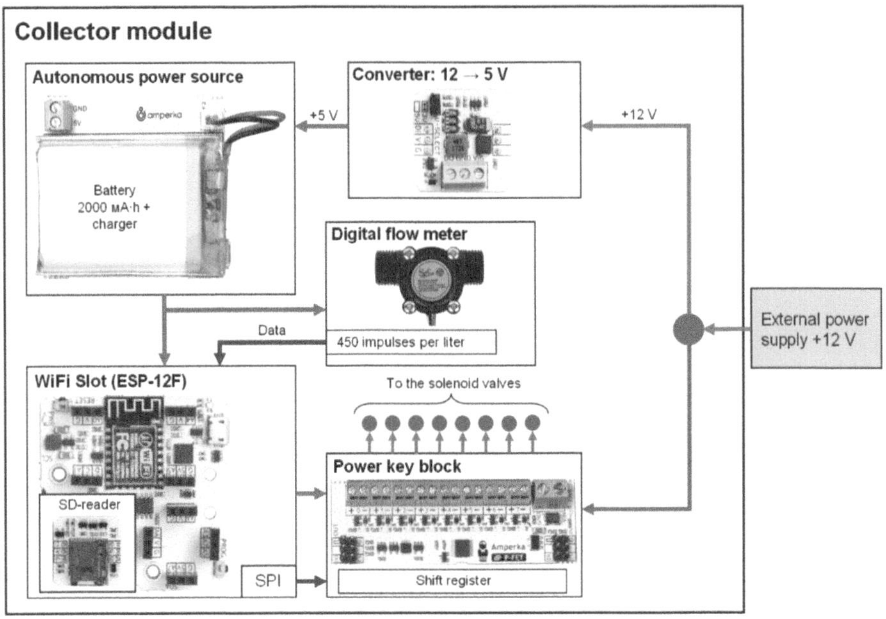

Fig. 7. Scheme of interaction of hardware of the collector module.

storage of XML file data on the SD card, supports interaction with the digital twin via the MQTT protocol, organizes the Web interface and the general logic of the device. When the ESP8266 collector module is initially turned on, it operates in access point mode and allows the user to connect to its network. After that, a web form can be opened in the browser at 192.168.0.1 for further configuration of the module and integration of its microclimate control system. The Web server, which operates on the microcontroller of the collector module, provides interactive irrigation control, schedule programming, and receiving summary data on task execution. The operation diagram of the built-in Web application is shown in Fig. 8. When updating Web form data, the server uses the status of irrigation lines, flow meter readings and timestamps, which are transmitted to the client's browser using AJAX technology.

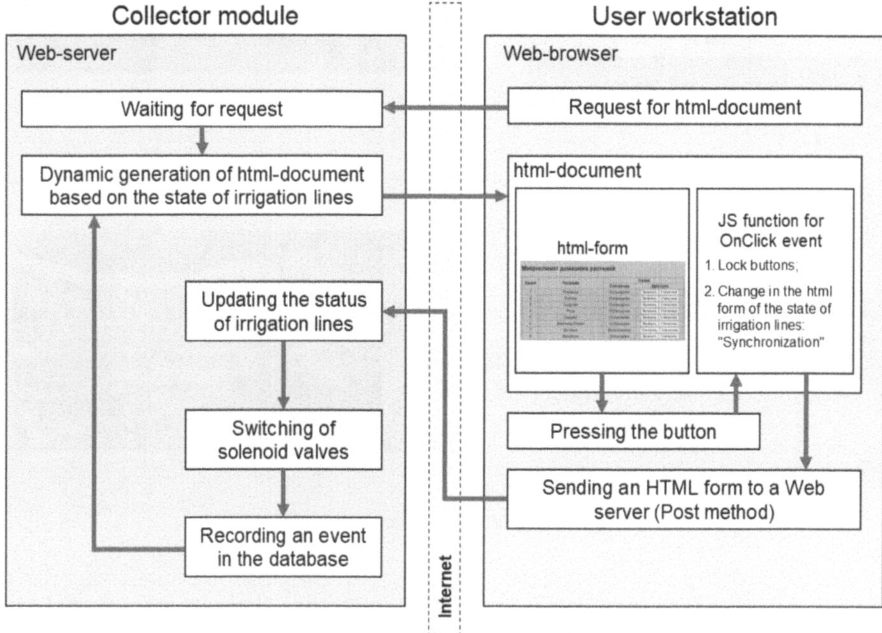

Fig. 8. Scheme of operation of the Web service of the collector module.

The data generated by the mechatronic module hardware during operation is formatted by the microcontroller software into XML files. To ensure autonomous operation of the device, files are first saved in the built-in flash memory. They are then published to the Internet in the background via an MQTT broker. The latter distributes data between subscriber services. Such subscribers can be either a digital twin of the collector module or other virtual services.

Figure 9 shows the testing process of the web application of the collector module for watering house plants.

The web form displays the correspondence between irrigation lines and plants, and also provides the user with buttons to turn on/off the liquid supply and display the history of plant irrigation. The program logic provides for sequential watering, so activation of any line leads to the shutdown of another, if it was used. During the process of switching valves, information about the selected button is sent to the server via a Post request. In this case, the user's browser blocks the elements of the Web form, and the irrigation line status displays the "Synchronization" mark. Receiving a Post-request by the server results in switching the water supply valves of the collector module, recording information about this event in the device memory, forming an XML container and sending it to the client

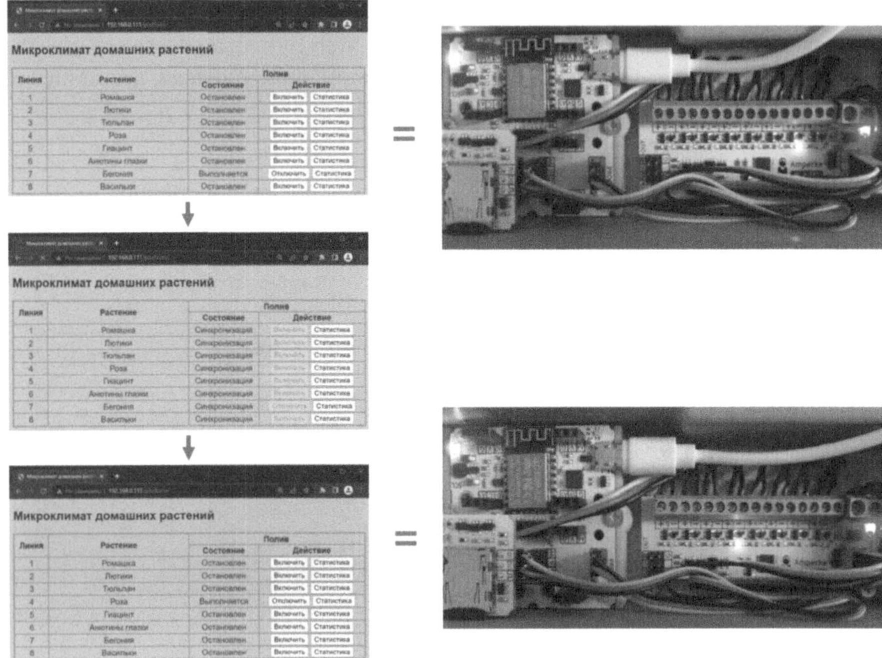

Fig. 9. Testing the collector module.

browser. After the Web form data is updated, its interactive elements are activated and again allow the user to manipulate the irrigation.

4 Conclusion

The Wi-Fi Slot embedded computing platform with the ESP8266 microcontroller has significant potential for increased performance and functionality. Its hardware resources can be scaled both through a wide range of functional blocks connected via SPI, and through connections to multiple Slot Expander microcontroller assemblies, which are interfaced via the I2C protocol. The type of power key block used in the collector module also makes it possible to significantly increase the number of plants being serviced. Virtual expansion of the shift register bit depth with serial SPI connection of such blocks makes it possible to increase the number of electromagnetic valves controlled by one microcontroller of the collector module.

The ESP8266 microcontroller of the collector module is capable of supporting the operation of the Web application. However, with the expansion of the range of functions of the software of the mechatronic module, it will be rational to transfer it to the virtual service of the digital twin, with which the user of the microclimate control system will interact first. The digital twin will be able to respond to a request for summary data independently. However, if an irrigation control command is received, the request will need to be forwarded to the physical device.

The development of microcontroller technologies, FPGA and special processors for vector and tensor calculations allows integrating virtually any data processing methods into measuring and executive devices. On the one hand, this complicates the internal structure of these devices and forces us to step beyond the operating system boundary when organizing network control complexes. On the other hand, the ability to transform data in intelligent devices facilitates the transition to a higher-order signal system, which significantly reduces the traffic of measuring network management, makes it possible to unify the approach to synchronizing devices in the process of data recording, eliminates the need for intermediate interfaces, overcome the limited resources of a single computing platform of a measuring complex and increase its speed.

References

1. Ganssle, J.: The Art of Designing Embedded Systems. Butterworth-Heinemann (2000)
2. Cai, J., Ran, F., Xu, M., Zhen, L.: Design of low-power 10 bit 40 MS/s pipelined ADC converter. Huazhong Keji Daxue Xuebao (Ziran Kexue Ban). J. Huazhong Univ. Sci. Technol. (Nat. Sci. Ed.) **38**(1), 61–64 (2010)
3. Chernysheva, N.S., Ionov, A.B.: Application of a multichannel radiation thermometer for increase in adequacy of non-contact temperature measurement results. In: International Conference of Young Specialists on Micro/Nanotechnologies and Electron Devices, EDM August, 2015, pp. 334–336 (2015). 7184557
4. Ammar, A., Fredj, H.B., Souani, C.: Accurate realtime motion estimation using optical flow on an embedded system. Electronics **10**(17), 2164 (2021)
5. Cooklev, T., Eidson, J.C., Pakdaman, A.: An implementation of IEEE 1588 over IEEE 802.11 b for synchronization of wireless local area network nodes. IEEE Trans. Instrum. Meas. **56**(5), 1632–1639 (2007)
6. Cochran, R., Marinescu, C.: Design and implementation of a PTP clock infrastructure for the Linux kernel. In: IEEE International Symposium on Precision Clock Synchronization for Measurement, Control and Communication, pp. 116–121. IEEE (2010)
7. Dolmatov, A.V., Gulyaev, P.Yu.: Thermal imaging complex with tracking function for joint research of microheterogeneous processes and macrokinetics of SHS phenomenon. J. Phys. Conf. Ser. **1333**(6), 062006 (2019). https://doi.org/10.1088/1742-6596/1333/6/062006
8. Dolmatov, A., Gulyaev, P., Milyukova, I.: Architecture of an intelligent network pyrometer for building information-measuring and mechatronic systems. Commun. Comput. Inf. Sci. **1526**, 265–274 (2022). https://doi.org/10.1007/978-3-030-94141-3_21
9. Godovnikov, E.A., Andreeva, E.G., Kovalev, V.Z., Usmanov, R.T.: Identification of parameters of power circuits pulse energy conversion systems of electromechanical equipment. J. Phys. Conf. Ser. **1260**(5), 052007 (2019). https://doi.org/10.1088/1742-6596/1260/5/052007
10. Chen, P., Yang, Z.: Understanding precision time protocol in today's Wi-Fi networks: a measurement study. In: 2021 USENIX Annual Technical Conference (USENIX ATC 21), pp. 597–610. USENIX Association (2021)

Potential Control Strategy of Mobile Robot in Dynamic Environment Based on Modified Bug Algorithm

Mikhail A. Kozhin[1] and Nikolay B. Filiomonov[1,2(✉)]

[1] Lomonosov Moscow State University, Leninskie Gory, 1/2, 119991 Moscow, Russia
nbfilimonov@mail.ru

[2] Bauman University, 2nd Baumanskaya Street, 5/1, 105005 Moscow, Russia

Abstract. This paper is dedicated to the development and study of a potential control strategy for a mobile robot operating in a partially known dynamic environment, based on a modified bug algorithm. Improvements to both the modified bug algorithm and the potential control strategy are proposed. The modified bug algorithm is used to generate a reference path for the mobile robot. The potential control strategy, which combines the modified bug algorithm, the artificial potential field method, and proactive control, is employed to correct the reference path. The developed algorithm was tested in the CoppeliaSim simulator using Python scripts.

Keywords: Mobile Robots · Dynamic Environment · Obstacle Avoidance · Artificial Potential Field · Bug Algorithm · Potential Control Strategy

1 Introduction

One of the most important challenges in modern robotics is the motion control of mobile robots (MRs) to ensure safe and efficient navigation toward a goal. In many applications, the operating environment is dynamic, containing moving obstacles (DOs). When all information about static obstacles is known but the dynamic obstacles are only partially observable, the environment is referred to as a partially known dynamic environment. In such settings, the robot's reaction speed to moving obstacles and its ability to act proactively are critical for safe and successful navigation.

This paper presents further development of a potential control strategy for a mobile robot operating in a partially known dynamic environment. The modified bug algorithm is used to generate a reference path for the robot, which is then corrected using the proposed potential control strategy to ensure collision-free motion.

2 Control of Mobile Robots in Obstacle-Rich Environments

The control of a mobile robot (MR) involves two main tasks: navigation and trajectory planning. The primary approaches to navigation include satellite-based navigation (e.g., GPS), radio beacons, and odometry. Satellite navigation is generally unsuitable

V. Jordan et al. (Eds.): HPCST 2024, CCIS 2919, pp. 202–211, 2026.
https://doi.org/10.1007/978-3-032-20325-0_16

for mobile robots, particularly indoors, because GPS signals may be unavailable and its accuracy is often lower than the size of the robot. Radio beacons are effective indoors but require pre-installed infrastructure and limit the robot's ability to choose alternative paths. Odometry, including visual odometry [1–3] and thermal-inertial odometry [4–6], is widely used for MR navigation. Visual odometry requires sufficient illumination and minimal visual obstructions such as fog or smoke, while thermal-inertial odometry is limited indoors due to small temperature gradients, resulting in few reliable landmarks. In this work, we focus on MRs equipped with ultrasonic sensors, which are commonly used in odometry.

Trajectory planning in environments with obstacles can be approached using several methods:

- **Graph-based methods** – including rapidly-exploring random trees (RRT) [7], Voronoi diagrams [8], visibility graphs [9], and probabilistic roadmaps [10]. These methods treat the robot's start, intermediate, and goal positions as graph nodes. While promising, they have limitations: RRT may not converge to an optimal solution and requires significant memory; Voronoi and visibility-based methods require map pre-processing; and probabilistic roadmaps lose efficiency in environments with narrow passages or bottlenecks.
- **Cellular decomposition methods** – such as Dijkstra, A*, and D* algorithms [11–13]. These methods represent the environment as an occupancy grid and apply graph-based pathfinding. However, they often produce non-smooth trajectories, necessitating additional smoothing algorithms [14, 15].
- **Intelligent algorithms** – including neural networks, genetic algorithms, and other heuristic approaches [16–19]. These methods are flexible but require powerful computational resources.
- **Potential field methods** – which are computationally efficient and naturally produce smooth trajectories [20]. In this approach, obstacles are modeled as repulsive forces, while the goal exerts an attractive force, guiding the robot toward the target while avoiding collisions. The main drawbacks are the possibility of local minima and difficulty handling dynamic obstacles.

3 Dynamic Model of Mobile Robot Motion

A two-wheeled MR equipped with ultrasonic proximity sensors is considered. The sensors visibility range ρ_{v} is assumed to be known.

Let vector $\mathbf{r} = (x, y)$ connecting the origin of coordinates with the center of mass (CM) of the robot (see Fig. 1) be a radius vector of the MR. The spatial orientation of the MR is the angle θ between the MR velocity vector $\mathbf{v}$ and Ox axis. Let also V be the MR's velocity value, Ω - the MR's angular velocity, J - moment of inertia of the robot when rotating around a vertical axis passing through it's CM.

Let $\mathbf{d}_0$ be a vector connecting the robot's CM with the closest obstacle point and $\mathbf{d}_{\mathrm{g}}$ be a vector connecting the robot's CM with the goal. Then the dynamics of the MR's

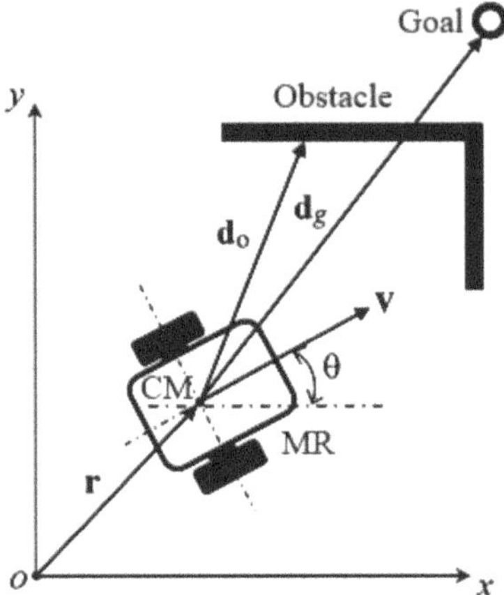

Fig. 1. An example of MR operating environment

motion is described by the following system of differential equations:

$$\begin{cases} \dot{x} = V\cos(\theta), \\ \dot{y} = V\sin(\theta), \\ \dot{\theta} = \Omega, \\ m\dot{V} = m_p a\Omega^2 + \frac{u_1 + u_2}{R}, \\ J\dot{\Omega} = m_p a\Omega V + u_1 - u_2, \end{cases}$$

where m and m_p are the robot's mass and the platform's mass correspondingly, a is a half of the distance between the robot's wheels, R is a robot's wheels radius.

4 Mobile Robot Navigation with Modified Bug Algorithm

Artificial potential field (APF) methods [20] are among the most promising approaches for MR motion control. In these methods, the robot and obstacles are modeled as negatively charged objects, while the goal is modeled as positively charged. As a result, the robot is repelled by obstacles and attracted to the goal. A key limitation of APF methods is the possibility of the MR becoming trapped in local minima, from which it cannot escape autonomously.

Classical bug algorithms [21] provide safe navigation in environments without dynamic obstacles. These algorithms move the MR along the gradient vector that leads to the shortest path toward the target in free space, bypassing obstacles as needed. Their main drawback is that the MR may need to traverse a large portion of the obstacles' perimeter, increasing path length.

The modified bug algorithm (MBA) [22–24] combines ideas from classical bug algorithms and APF methods:

- in free space, the MR moves toward the goal along a straight line;
- at each point, the MR's velocity vector is determined by the superposition of artificial potential fields generated by obstacles and the goal.

When an obstacle is detected within a radius ρ_0 in the direction of motion, the robot initiates an obstacle-bypass maneuver. The bypass is considered complete if any of the following conditions are met:

- the robot has moved closer to the goal;
- there are no obstacles within the visibility range along the direction of the target;
- the angle between the vectors $\mathbf{d}_0$ to obstacle and $\mathbf{d}_g$ to the goal is not acute.

Additionally, the obstacle-bypass maneuver is terminated if no obstacle is detected within the visibility range.

The artificial force of attraction to the goal $\mathbf{F}_{att}(\mathbf{r})$ has the following form:

$$\mathbf{F}_{att}(\mathbf{r}) = k_a \frac{\mathbf{d}_g(\mathbf{r})}{\|\mathbf{d}_g(\mathbf{r})\|},$$

where k_a is a positive constant.

In common MBA the artificial forces of repulsion from obstacles $\mathbf{F}_{rep}(\mathbf{r})$ has the following form:

$$\mathbf{F}_{rep}(\mathbf{r}) = \begin{cases} -k_r\left(\frac{1}{\|\mathbf{d}_0(\mathbf{r})\|} - \frac{1}{\rho_r}\right)\frac{\mathbf{d}_0(\mathbf{r})}{\|\mathbf{d}_0(\mathbf{r})\|^3}, & \|\mathbf{d}_0(\mathbf{r})\| \le \rho_r, \\ 0, & \|\mathbf{d}_0(\mathbf{r})\| > \rho_r, \end{cases}$$

where k_r is a positive constant and ρ_r is a width of the repulsive force action zone.

However, this form of the repulsive potential together with the finite time step can cause trajectory smoothness problems. Thus the following form of the $\mathbf{F}_{rep}(\mathbf{r})$ is proposed:

$$\mathbf{F}_{rep}(\mathbf{r}) = \begin{cases} -k_r\left(\frac{1}{\|\mathbf{d}_0(\mathbf{r})\|} - \frac{1}{\rho_r}\right)\frac{\mathbf{d}_0(\mathbf{r})}{\|\mathbf{d}_0(\mathbf{r})\|^3}, & \|\mathbf{d}_0(\mathbf{r})\| \le \rho_v, \\ 0, & \|\mathbf{d}_0(\mathbf{r})\| > \rho_v, \end{cases}$$

where $\rho_r < \rho_v$.

As a result, the robot will be forced to bypass the obstacle until the completing conditions are met.

The artificial tangential force $\mathbf{F}_{tan}(\mathbf{r})$ that helps to bypass obstacles has the following form:

$$\mathbf{F}^{\pm}_{tan}(\mathbf{r}) = \frac{k_a}{\|\mathbf{d}_0(\mathbf{r})\|}\left(\pm d_{0_2}, \mp d_{0_1}\right),$$

where $\mathbf{F}^{+}_{tan}(\mathbf{r})$ corresponds to bypass from the right and $\mathbf{F}^{-}_{tan}(\mathbf{r})$ - from the left.

In this case the artificial resultant force $\mathbf{u}(\mathbf{r})$ will be the following:

$$\mathbf{u}(\mathbf{r}) = \begin{cases} \frac{\mathbf{F}_{att}(\mathbf{r})}{\|\mathbf{F}_{att}(\mathbf{r})\|}, & \text{in free space;} \\ \frac{\mathbf{F}_{rep}(\mathbf{r})+\mathbf{F}_{tan}(\mathbf{r})}{\|\mathbf{F}_{rep}(\mathbf{r})+\mathbf{F}_{tan}(\mathbf{r})\|}, & \text{when bypassing obstacle.} \end{cases}$$

In this paper the MBA is used to create a reference MR's motion path. The created reference path and an example of the MR's operating environment is presented on the Fig. 2. The reference path is created with the following constants: $\rho_r = 0.5$, $\rho_0 = 1$, $\rho_v = 1$, $k_a = 1$, $k_r = 1$. However, the reference path does not take into account dynamic obstacles, so it needs to be corrected. To do this the potential mobile robot control strategy will be used.

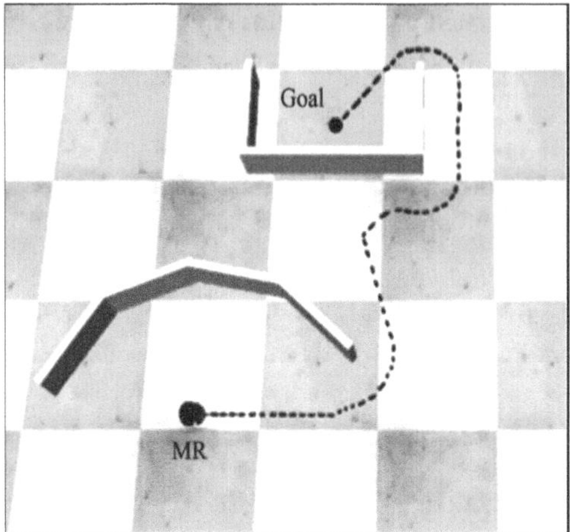

Fig. 2. An example of the MR's operating environment and a created reference path

5 Mobile Robot Control Using Potential Fields

The potential control strategy (PCS) is based on three methods [24]: the modified bug algorithm, artificial potential field methods, and proactive control. Two motion modes of the mobile robot are distinguished:

- motion in the absence of dynamic obstacles within the visibility range;
- motion when dynamic obstacles are present within the visibility range.

In the motion mode without dynamic obstacles, the MR moves toward the goal along the reference path.

In the motion mode with DO, the MR corrects the reference path to avoid collisions using the proactive control algorithm. This algorithm involves three key operations:

- predicting potential collisions between the MR and DO;
- selecting an appropriate attraction point for the MR to avoid the DO;
- moving the MR while adjusting the reference path to safely avoid the DO.

Prediction of collision of the MR and DO is carried out by finding the DO velocity vector $\mathbf{v}_0$ using proximity sensors data:

$$\mathbf{v}_0 = \frac{\mathbf{d}_2 - \mathbf{d}_1}{\Delta t} + \mathbf{v},$$

where $\mathbf{d}_1$ and $\mathbf{d}_2$ are vectors which connect the CM of the robot with the closest DO point in a different time moments that are separated by a small time Δt (see Fig. 3).

Let's define the collision vector as $\mathbf{d}_\text{m} = \mathbf{d}_2 + (\mathbf{v}_0 - \mathbf{v})\,t$, where t - some point in time. If $\exists t \in [0,\ T]$:

$$\|\mathbf{d}_\text{m}\| < \rho_\text{s},$$

then the collision is considered predicted in the next T seconds. Here ρ_s is a minimum safe distance from the robot's CM to the closest obstacle point.

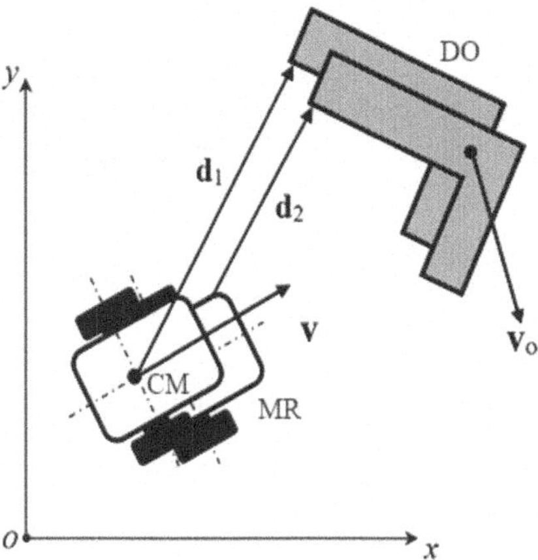

Fig. 3. An example of the MR's and DO's movements

Selection of attraction point for MR to avoid the DO is carried out by a sequential sorting out of the reference path points starting from the point at which the robot left the reference path. The points are sorted out until the next one is further from the robot than the previous one.

For the first chosen point vector $\mathbf{d}_t$ from the MR's CM is calculated and stored (see Fig. 4).

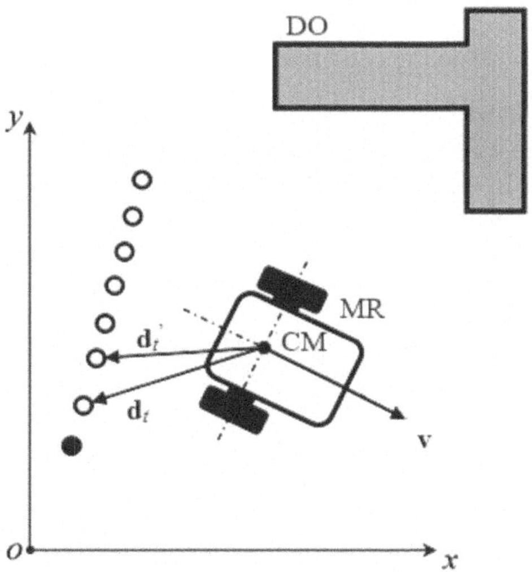

Fig. 4. MR leaves the reference path

In case the following condition is met for the next reference path point with vector $\mathbf{d}_t'$: $\left\|\mathbf{d}_t'\right\| \leq \|\mathbf{d}_t\|$, then a new vector $\mathbf{d}_t = \mathbf{d}_t'$ is stored. This cycle is repeated until the condition $\|\mathbf{d}'_t\| \leq \|\mathbf{d}_t\|$ is met. The last stored point is called the MR's attraction point when avoiding the DO.

Movement with the reference path correction for MR to avoid the DO is carried out in the direction of the artificial resultant force, the components of which is formed by the following rules:

1. If at least one of the following conditions is met:

- collision is predicted;
- DO is located in the direction of the MR's movement;
- MR is moving away from the attraction point.

Then MR is moving under the influence of the artificial tangential and repulsive from DO forces:

$$\mathbf{F}^s_{\tan}(\mathbf{r}) = \frac{k_a}{\|\mathbf{d}_2(\mathbf{r})\|}\left(\pm d_{2_2}, \mp d_{2_1}\right),$$

$$\mathbf{F}_{\text{rep}}(\mathbf{r}) = \begin{cases} -k_r\left(\frac{1}{\|\mathbf{d}_2(\mathbf{r})\|} - \frac{1}{\rho_r}\right)\frac{\mathbf{d}_2(\mathbf{r})}{\|\mathbf{d}_2(\mathbf{r})\|^3}, & \|\mathbf{d}_2(\mathbf{r})\| \leq \rho_r, \\ 0, & \|\mathbf{d}_2(\mathbf{r})\| > \rho_r, \end{cases}$$

where the direction of the DO bypass s in $\mathbf{F}^s_{\tan}(\mathbf{r})$ is defined as follows:

$$s = \text{sgn}(-\cos(\mathbf{d}_m, \mathbf{v})\left(\mathbf{v}_0 \mathbf{d}_{2_1} - \mathbf{v}_1 \mathbf{d}_{2_0}\right).$$

In the original PCS the DO bypass direction was the same as the one in reference path. However, this could lead to a collision with DO.

3. If at least one of the following conditions is met:

- collision is predicted;
- MR is moving in the direction of the static obstacle;
- static obstacle is in the direction of the MR's attraction point and on way to it;

Then the MR is moving under the influence of the artificial repulsive from the static obstacle force:

$$\mathbf{F}_{\text{rep}}(\mathbf{r}) = \begin{cases} -k_r\left(\frac{1}{\|\mathbf{d}_0(\mathbf{r})\|} - \frac{1}{\rho_r}\right)\frac{\mathbf{d}_0(\mathbf{r})}{\|\mathbf{d}_0(\mathbf{r})\|^3}, & \|\mathbf{d}_0(\mathbf{r})\| \leq \rho_v, \\ 0, & \|\mathbf{d}_0(\mathbf{r})\| > \rho_v. \end{cases}$$

Moreover, if MR is not under the influence of a tangential force from the DO then the following tangential force from the static obstacle arises:

$$\mathbf{F}^{\pm}_{\tan}(\mathbf{r}) = \frac{k_a}{\|\mathbf{d}_0(\mathbf{r})\|}\left(\pm d_{0_2}, \mp d_{0_1}\right),$$

where the bypass direction is the same as the one in the reference path.

Computer testing of the proposed PCS (see Fig. 5) was performed with the following constants: $\rho_r = 0.5$, $\rho_0 = 1$, $\rho_v = 1$, $\rho_s = 0.25$, $k_a = 1$, $k_r = 1$, $T = 5$.

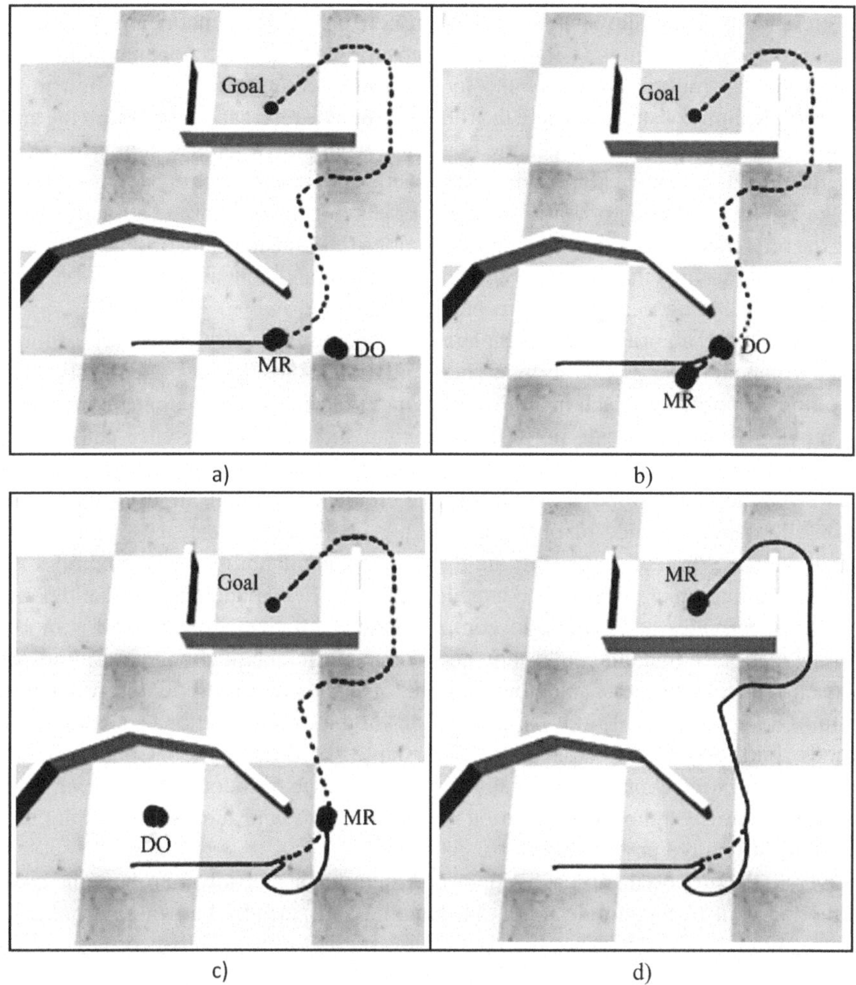

Fig. 5. Reference path correction during dynamic obstacle avoidance: (a) Detection of a dynamic obstacle by the mobile robot's proximity sensors, marking the switch of motion modes; (b) Dynamic obstacle avoidance using the proactive control algorithm; (c) Return to the initial reference path via the selected attraction point after successful obstacle avoidance; (d) Mobile robot reaching the goal.

6 Conclusion

The potential control strategy for mobile robots operating in partially known dynamic environments has been presented and analyzed. This strategy integrates the modified bug algorithm for reference path generation with artificial potential field methods and a proactive control approach for real-time path correction. The modified bug algorithm provides a reliable baseline trajectory by combining the efficiency of straight-line motion toward the goal with obstacle avoidance capabilities. The potential control strategy enhances this reference path by allowing the mobile robot to respond dynamically to moving obstacles detected within its visibility range, ensuring safe and efficient navigation.

Two distinct motion modes of the mobile robot were considered: motion in the absence of dynamic obstacles, where the robot follows the reference path toward the goal, and motion in the presence of dynamic obstacles, where the reference path is corrected using the proactive control algorithm. The proactive control algorithm operates through three key steps: predicting potential collisions, selecting an appropriate attraction point along the reference path, and guiding the robot along the adjusted path to avoid collisions. This approach ensures that the robot maintains smooth and collision-free motion while dynamically adapting to changes in its environment.

The developed algorithms were implemented and tested in the CoppeliaSim simulation environment using Python scripts. The simulation results demonstrate the effectiveness of the proposed approach in a range of dynamic scenarios, including environments with moving obstacles of varying trajectories and speeds. The reference path correction mechanism successfully prevented collisions, minimized detours around obstacles, and allowed the mobile robot to reach the goal efficiently. Visualizations of the robot's motion show the sequence of obstacle detection, path adjustment, and return to the original reference trajectory, confirming the practical applicability of the method.

The combination of the modified bug algorithm, artificial potential fields, and proactive control provides a flexible and computationally efficient framework for mobile robot navigation in dynamic environments. Compared to classical bug algorithms, the approach reduces unnecessary detours around obstacles and prevents the robot from becoming trapped in local minima, a common limitation of traditional potential field methods. Furthermore, the strategy can be adapted to different robot platforms and sensor configurations, making it suitable for a wide range of indoor and outdoor applications, including warehouse automation, service robotics, and autonomous exploration in partially structured environments.

Overall, the proposed approach represents a significant step toward robust and adaptive motion control for mobile robots in dynamic and partially known environments. By integrating reliable reference path planning with real-time adaptive corrections, the method ensures both safety and efficiency, and it can serve as a foundation for further research and development in autonomous mobile robotics.

References

1. Qin, T., Li, P., Shen, S.: VINS-Mono: a robust and versatile monocular visual-inertial state estimator. IEEE Trans. Robot. **34**(4), 1004–1020 (2018)
2. Sun, K., et al.: Robust stereo visual inertial odometry for fast autonomous flight. IEEE Robot. Autom. Lett. **3**(2), 965–972 (2018)
3. Bloesch, M., Omari, S., Hutter, M., Siegwart, R.: Robust visual inertial odometry using a direct EKF-based approach. In: 2015 IEEE/RSJ Conference on Intelligent Robots and Systems (IROS), pp. 298–304 (2015)
4. Saputra, M.R.U., et al.: DeepTIO: a deep thermal-inertial odometry with visual hallucination. IEEE Robot. Autom. Lett. **5**(2) (2020)
5. Khattak, S., Papachristos, C., Alexis, K.: Keyframe-based thermal-inertial odometry. J. Field Robot. **37**(4), 552–579 (2020)

6. Zhao, S., Wang, P., Zhang, H., Fang, Z., Scherer, S.: TP-TIO: a robust thermal-inertial odometry with deep thermalpoint. In: 2020 IEEE/RSJ International Conference on Intelligent Robots and Systems (IROS)
7. LaValle, S.M., Kuffner, J.J.: Rapidly-exploring random trees: progress and prospects. In: Workshop on the Algorithmic Foundations of Robotics, pp. 293–308 (2000)
8. Guibas, L.J., Knuth, D.E., Sharir, M.: Randomized incremental construction of Delaunay and Voronoi diagrams. Algorithmica **7**(1), 381–413 (1992)
9. De Berg, M., Cheong, O., Van Kreveld, M., Overmars, M.: Computational Geometry: Algorithms and Applications, 3rd edn. Springer, Heidelberg (2008). https://doi.org/10.1007/978-3-540-77974-2
10. Kedem, K., Sharir, M.: An efficient motion planning algorithm for a convex rigid polygonal object in 2-dimensional polygonal space. Discrete Comput. Geom. **5**(1), 43–75 (1990)
11. Kazakov, K.A., Semenov, V.A.: Obzor sovremennykh metodov planirovaniya puti [Review of modern methods of path planning]. Tr. ISP RAN **28**(4), 241–294 (2016)
12. Hart, P.E., Nilsson, N.J., Raphael, B.A.: Formal basis for the heuristic determination of minimum cost paths. IEEE Trans. Syst. Sci. Cybern. **2**, 100–107 (1968)
13. Stentz, A.: Optimal and efficient path planning for partially known environments. In: Hebert, M.H., Thorpe, C., Stentz, A. (eds.) Intelligent Unmanned Ground Vehicles. SECS, vol. 388, pp. 203–220. Springer, Boston (1997). https://doi.org/10.1007/978-1-4615-6325-9_11
14. Zhang, X., Wylie, B., Oscar, C., Moore, C.A.: Time-optimal and collision-free path planning for dual-manipulator 3D printer. IEEE Trans. Syst. Man Cybern. Syst., 2389–2396 (2020)
15. Pshikhopov, V.Kh., Medvedev, M.Yu., Kostyukov, V.A., Khusseyn, F., Kadim, A.: Algoritmy planirovaniya traektoriy v dvumernoy srede s prepyatstviyami [Algorithms for trajectory planning in a two-dimensional environment with obstacles]. Informatika i Avtomatizatsiya **21**(3), 459–492 (2022)
16. Güzel, M., Ajabshir, V., Nattharith, P., Gezer, E., Can, S.: A novel framework for multi-agent systems using a decentralized strategy. Robotica **37**(4), 691–707 (2019)
17. Medvedev, M., Pshikhopov, V.: Path planning of mobile robot group based on neural networks. In: Fujita, H., Fournier-Viger, P., Ali, M., Sasaki, J. (eds.) IEA/AIE 2020. LNCS, vol. 12144, pp. 51–62. Springer, Cham (2020). https://doi.org/10.1007/978-3-030-55789-8_5
18. Wang, Y., Cheng, L., Hou, Z.-G., Yu, J., Tan, M.: Optimal formation of multirobot systems based on a recurrent neural network. IEEE Trans. Neural Netw. Learn. Syst. **27**(2), 322–333 (2016)
19. Gayduk, A.R., Mart'yanov, O.V., Medvedev, M.Yu., Pshikhopov, V.Kh., Khamdan, N., Farkhud, A.: Neyrosetevaya sistema upravleniya gruppoy robotov v neopredelennoy dvumernoy srede [Neural network control system for a group of robots in an indefinite two-dimensional environment]. Mekhatronika Avtomatizatsiya Upravlenie **21**(8), 470–479 (2020)
20. Khatib, O.: Real-time obstacle avoidance for manipulators and mobile robots. In: IEEE International Conference on Robotics and Automation, St. Louis, MO, USA, pp. 500–505 (1985)
21. Ng, J., Braunl, Th.: Performance comparison of bug navigation algorithms. J. Intell. Robot. Syst. **50**(1), 73–84 (2007)
22. Filimonov, A.B., Filimonov, N.B., Barashkov, A.A.: Issues of motion control of mobile robots based on the potential guidance method. Mechatron. Autom. Control **20**(11), 677–685 (2019). (in Russian)
23. Filimonov, A.B., Filimonov, N.B., Kozhin, M.A.: Analysis of modified bug algorithm in mobile robot local navigation problem. Mechatron. Autom. Robot. **7**, 4–8 (2021). (in Russian)
24. Kozhin, M.A.: Potential control strategy of mobile robot in dynamic environment. Mechatron. Autom. Robot. **12**, 26–32 (2023). (in Russian)

Computer-Assisted Measuring System for Study of Aluminum-Nickel Alloys

Vladimir Malikov[1](✉), Sergey Voinash[2], Irina Vornacheva[3], Vladimir Chaplygin[3], Ilia Kozlitin[3], Denis Fadeev[4], and Alexey Ishkov[5]

[1] Altai State Pedagogical University, Molodezhnaya Avenue 55, 656031 Barnaul, Russia
osys11@gmail.com

[2] Peoples' Friendship University of Russia named after Patrice Lumumba, 6 Miklukho-Maklaya Street, 117198 Moscow, Russia

[3] South-West State University, 50 let Oktyabrya Street, 94, 305040 Kursk, Russia

[4] Altai State University, Lenina Avenue 61, 656057 Barnaul, Russia

[5] Altai State Agricultural University, Krasnoarmeiskiy Avenue 98, 656049 Barnaul, Russia

Abstract. This work introduces a hardware and software system, built on the Raspberry Pi microcomputer and powered by Python, designed for the analysis of aluminum-nickel alloys. The study also provides a rationale for the relevance of investigating aluminum alloys with nickel additives. These alloys have high strength, low density, and high heat, wear and corrosion resistance. The paper describes a modernization of eddy-current transducer and complex making it possible to research the parts made of composite materials based on nickel and aluminum. Here, the results of the study of nickel and aluminum alloy samples made of powder intermetallides using spark plasma sintering with utilization of eddy current testing are presented.

Keywords: Microcomputer · Python · Eddy-Current transducer · Aluminum-Nickel Alloys · Hardware-Software Complex

1 Introduction

Intermetallic compounds in transition metal alloys like aluminum-nickel are increasingly studied for their unique combination of high strength, low density, and exceptional resistance to heat, wear, and corrosion [1–3].

Nickel aluminides are advanced materials for operation at high temperatures. They should demonstrate high strength, heat and fatigue resistance, high resistance to corrosion and oxidation, as well as sufficient ductility and appropriate technological properties [4–7]. Unlike traditional industrial alloys, nickel aluminides are characterized by lower density, high melting point and unique mechanical properties [8, 9].

The properties of Ni_3Al at elevated temperatures make it an ideal material for the production of turbine blades, gas turbine disks and diesel engine rotors. In addition, its high resistance to oxidation allows to use it in the production of glass-melting molds and clamping devices for work in high-temperature furnaces.

V. Jordan et al. (Eds.): HPCST 2024, CCIS 2919, pp. 212–225, 2026.
https://doi.org/10.1007/978-3-032-20325-0_17

However, parts made of this material require constant quality control to avoid possible damaging.

The main advantages of the eddy-current test of conducting research without the need for contact between the sensor and the test object, which plays an important role when choosing a research method.

2 FEM Analysis of Eddy Currents in Clamp-On Transducers

A critical objective in optimizing eddy current transducer (ECT) parameters is the numerical simulation of eddy current distribution within the test object. This study employs the FEM, a computational technique for obtaining approximate solutions to boundary value problems. The foundational principles of FEM involve two key steps: 1) the spatial discretization of the domain into a finite set of elements (meshing), and 2) the piecewise polynomial approximation of the governing field functions within these elements. A finite-element model developed to address the ECT optimization problem is presented in Fig. 1.

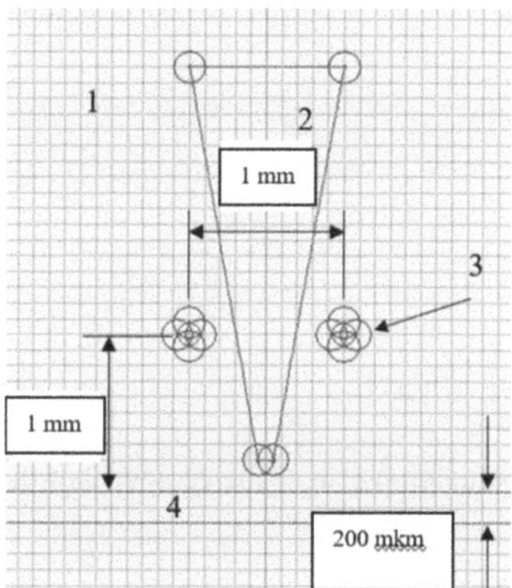

Fig. 1. Geometry of the problem.

The second stage involved assigning the material properties—namely, the electrical and magnetic parameters for the eddy current transducer, the test object, and the surrounding medium—to their corresponding domains within the geometric model (Table 1). The core material was specified as T35 ferrite, chosen due to its superior relative magnetic permeability of $\mu = 500$ at a frequency of 25 MHz, a characteristic that distinguishes it from alternative ferrite grades.

During the third stage, the geometric model is discretized into a finite element mesh. The selection of the element size is a critical step, governed by the need for numerical accuracy—specific to the problem's physics—and constrained by the available computational capacity. The final mesh configuration is illustrated in Fig. 2.

At the fourth stage, the equations for the magnetic potential, the scalar electric potential in each cell of the discretization grid were calculated.

Table 1. Electrical and magnetic parameters of the regions of the ECT model, the object of study and the environment.

Region	Substance	μ	σ, MSm/m	*j*, A
1	air	1	0	0
2	ferrite (T35)	500	$5 * 10^{-6}$	0
3	copper	1	57	0.02
4	copper	1	57	0

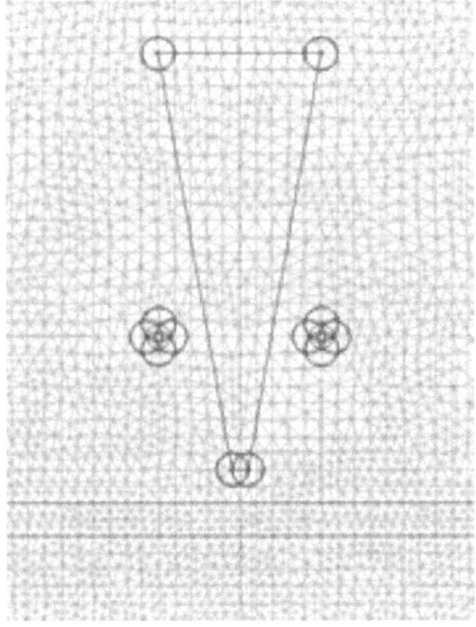

Fig. 2. Discretization grid of the geometry of the problem.

Modeling the distribution of eddy currents in a conductive object comes down to solving the problem of calculating the magnetic field of alternating currents.

The eddy current density ($j = -i\omega\gamma A$) (Fig. 3) are assumed to be constant. To describe the complete picture of the field, a system consisting of equations of the form is solved, the dimension of which is equal to the number of finite elements.

In accordance with the constructed theoretical model, a picture of the density of eddy currents in the object of study induced by an eddy current transducer without a core (Fig. 3. a) and with a ferrite core (Fig. 3. b) was obtained.

by an eddy current transducer.

Figure 4 shows results of modeling radial distribution of eddy current density in an object when an eddy-current current-carrying device interacts with it. The current density in the film was estimated at a depth of 1 μm, and the specific conductivity 57 MS/m. Modeling was performed both with a ferrite core (Fig. 4.1) and without a core (Fig. 4.2).

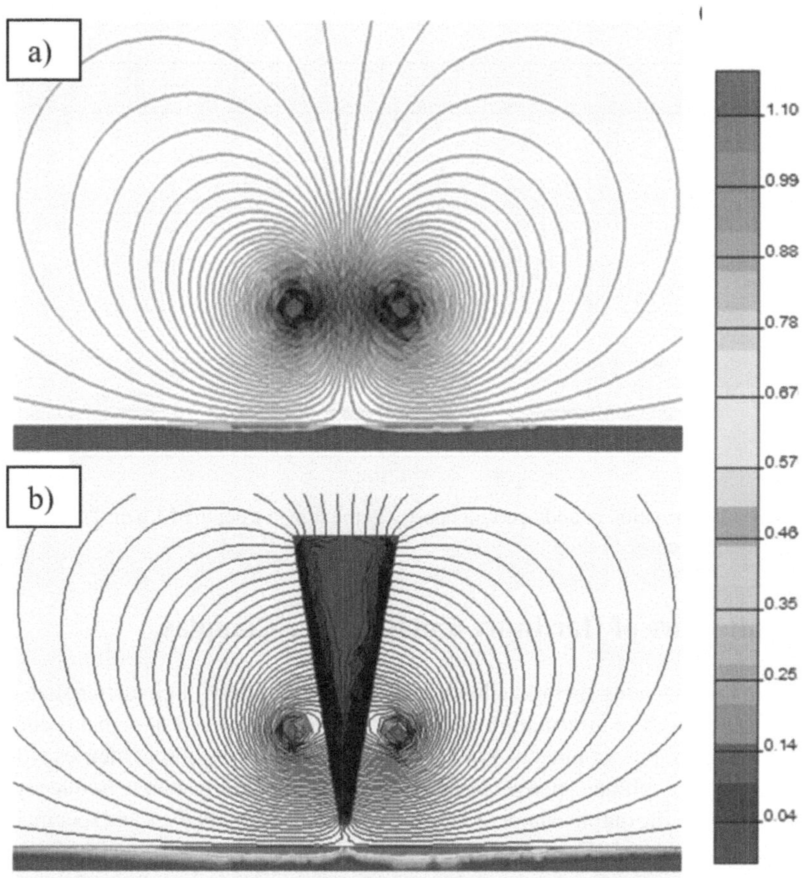

Fig. 3. Results of modeling the distribution of eddy currents in a thin metal film induced

Analysis of the simulation data (Fig. 4) indicates that the incorporation of a ferrite core resulted in a more than twofold increase in the maximum eddy current density within the film.

Modeling results established that inserting a ferromagnetic core into the coil significantly increases its inductance, attributable to the core's substantially higher magnetic permeability compared to air. Furthermore, shaping the core allows for control over the configuration of the magnetic flux generated by the coil, enabling the formation of a field with a specific geometry to enhance measurement resolution.

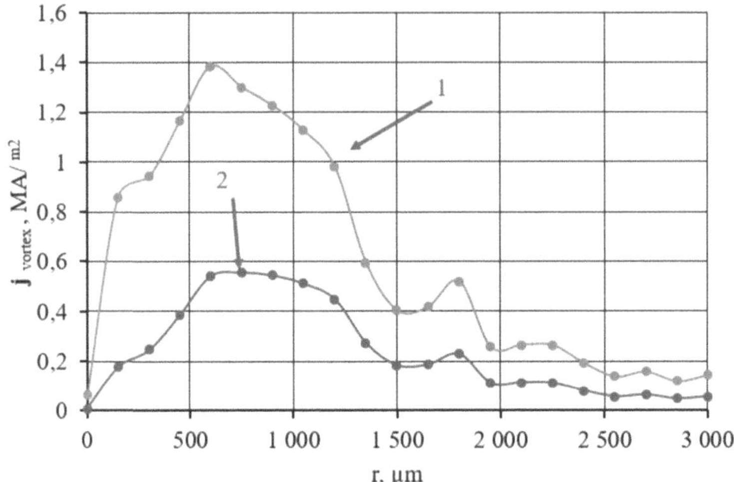

Fig. 4. Radial distribution of eddy current density in the film at a depth of 1 μm: 1) with a ferrite core), 2) without a core.

3 Architecture of Hardware and Software Complex

The eddy-current testing method allows to determine a variety of characteristics of the object under study. The information content of an electrical circuit or signal is encapsulated in its measurable parameters. For a transducer, the primary output metrics typically include voltage amplitude, phase, resistance, reactance, and resonant frequency. The selection of a specific output characteristic is dictated by the objectives and specifications of the measurement task. Furthermore, the methodological approach can be categorized based on its dimensionality: single-parameter, two-parameter, or multi-parameter analysis, each applicable to distinct measurement scenarios and complexity requirements [10, 11].

Based on empirical data from prior research, an eddy-current transducer was engineered to address the specified problem. The transducer features a 4.3-mm conical ferrite core with base and top diameters of 1.5 mm and 0.1 mm, respectively, and possesses an initial magnetic permeability of 50,000. The sensing winding is positioned at the conical tip, while the excitation winding is located at its center.

To perform this work, the study was carried out using the developed hardware and software complex based on a miniature eddy-current transducer (Fig. 5). The developed complex consists of four main parts: ECT, positioning system, and hardware and software components.

A dedicated signal generation system was developed to drive the excitation winding of the eddy-current transducer and generate its electromagnetic field. Integrated into a hardware-software complex, the system's software control is implemented via a generator module based on the AD9850 integrated circuit.

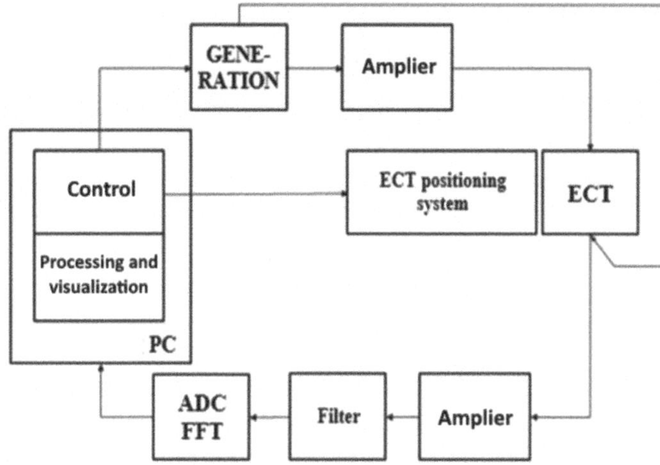

Fig. 5. Block diagram of the complex.

This microcircuit utilizes Direct Digital Synthesis (DDS) technology, which combines a high-speed digital-to-analog converter and a comparator to synthesize a programmable frequency and phase signal. When synchronized with a precise reference clock, the AD9850 generates a stable, low-noise analog sine wave output.

The module is characterized by an output frequency range of 1 to 40 MHz, high stability in frequency and amplitude, a low noise level, and a simplified control interface using a serial byte load format. It operates from a 3.3 V or 5 V supply, features low power consumption of 380 mW, and has compact dimensions of 3 $\times$ 4 cm.

The principal limitations of the developed generation module are its fixed output signal amplitude and limited output power. To mitigate these constraints, an external amplification stage with programmable gain was integrated into the system.

Module control is implemented via the Arduino hardware computing platform. This platform comprises two core components: a physical input/output board and a cross-platform integrated development environment (IDE) based on the Processing/Wiring language framework.

Arduino modules are commercially available in various form factors, which differ not only in physical dimensions and pin count but also in their integrated hardware specifications, including the model of the microcontroller unit (MCU), system clock frequency, and the capacity of flash memory and random-access memory (RAM).

Arduino Nano was chosen out of the four main form factors of microcircuits. This choice was due to its maximum compactness (to connect to a PC, Nano does not require an adapter) and low power consumption. Arduino is based on Atmel ATMEGA8 and ATMEGA168 microcontrollers. The module circuits are published under a Creative Commons license, which makes it possible to freely use this microcircuit in the own developments.

The Nano hardware platform is architected around the ATmega328 microcontroller. This microcontroller integrates 32 KB of Flash memory for program storage and is supplemented by 2 KB of Static Random-Access Memory (SRAM). The platform provides eight analog input channels, each featuring a 10-bit analog-to-digital converter (ADC). In its standard configuration, the analog input voltage range is defined from ground to a reference level of 5 V.

Figure 6 shows the connection diagram of the AD9850 module to Arduino. Nano. The module power contacts VCC and GND are connected to the 3.3 V and GND contacts on the Arduino, respectively, and the contacts responsible for the serial command transmission W_CLK, FU_UD, DATA, and RESET are connected to the Arduino contacts A1, A2, A3, and A4, respectively.

4 Cartesian Kinematic-Based Positioning System

A Cartesian robotic positioning system was employed to facilitate the precise movement of the eddy-current transducer (ECT) over the surface of the test object. This system operates on a three-axis Cartesian coordinate framework.

Within this configuration, the test object is mounted on a platform that translates along the Y-axis. The ECT probe, secured in its holder, is independently actuated along the X and Z axes, enabling three-dimensional positioning relative to the sample.

The ECT is integrated directly with this sample positioning system. Its operational principle is based on interaction with the test object through a generated electromagnetic field..

Figure 7 shows interface of software window of the complex. In this window, such parameters for the study are set: the length of the measurement X, the number of measured lines, and the measurement speed.

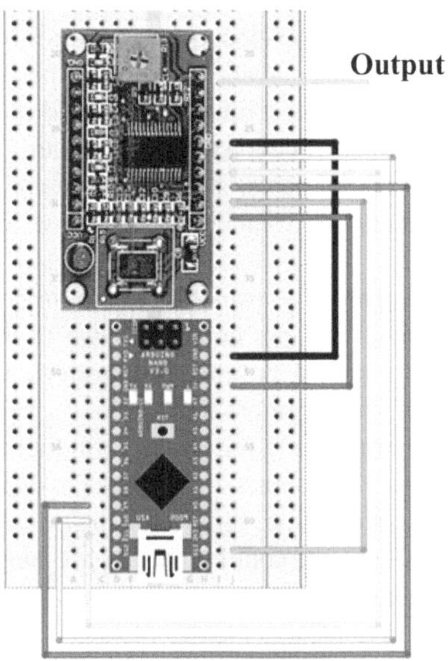

Fig. 6. Connection diagram of the AD 9850 module to Arduino Nano in the serial command transmission mode.

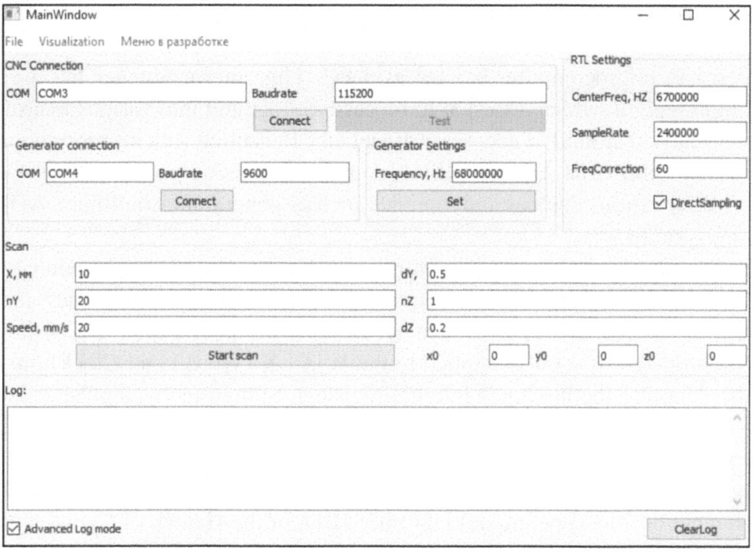

Fig. 7. Software graphical interface window.

The positioning system features a maximum scan area of 22 × 22 cm, with a maximum traverse velocity of 180 mm/s and a positioning accuracy of 100 μm.

A dedicated 32-bit processor motherboard governs the stepper motor operation. This motherboard interfaces with a host PC via a Universal Serial Bus (USB) for command reception and integrates HR4988 stepper motor drivers, which supply the requisite voltage and current to the motors. The system firmware is based on the open-source Marlin codebase, enabling operational control through standard G-code commands.

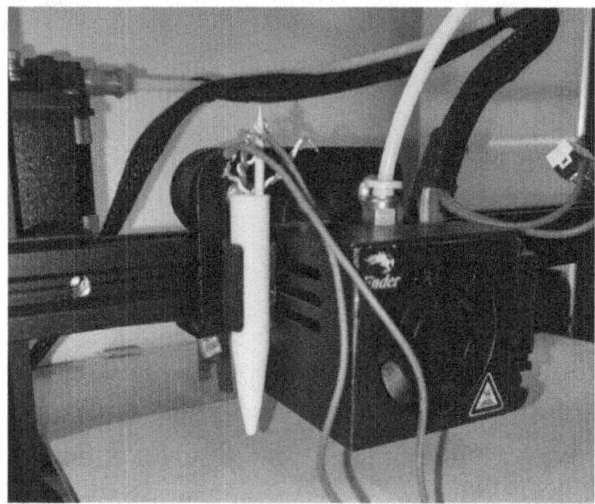

Fig. 8. ECT positioning system.

Figure 8 shows the eddy-current transducer location on the positioning system.

5 Digitization and Computerized Processing Data

A Raspberry Pi microcomputer is used as a PC. This microcomputer has a compact size and light weight, which allows it to be easily integrated into various measurement systems and carried around, if necessary. Raspberry Pi is fitted with a various connection interfaces including USB, HDMI, GPIO (common inputs/outputs), making it easy to integrate with a various devices and modules such as generators, amplifiers, ADCs, and positioning systems.

In serial mode, the Arduino microcontroller configures the AD9850 module by transmitting a 40-bit data sequence. This sequence, which defines the frequency and phase parameters of the output signal, is sent serially via the DATA line. The write operation is synchronized by a pulse applied to the W_CLK (Word Load Clock) pin, which latches the data into the module's internal registers. Subsequently, a pulse is issued to the FQ_UD (Frequency Update) pin. This final command commits the new parameters to the active synthesizer, initiating signal generation with the specified characteristics.

A digital USB oscilloscope with a built-in ADC was used to receive data (Fig. 9). A four-channel oscilloscope, model DSO-6254BD, of the Hantek of China, with a frequency of 1 GHz and a built-in signal generator with a frequency of up to 25 MHz, was chosen. This oscilloscope has a USB port via which the device can be connected to a computer.

Fig. 9. USB-oscilloscope DSO-6052BE.

The DSO 6052BE digital storage oscilloscope is characterized by the following key specifications and capabilities: it features two independent input channels with a system bandwidth of 50 MHz and is equipped with an Autoset function for automated signal conditioning. The instrument achieves a real-time sampling rate of 150 MS/s in single-channel operation. It supports 32 types of automated parametric measurements and offers integration with the LabVIEW® software environment. For custom application development, a programming library compatible with Visual C and Visual Basic is provided. The device also includes built-in mathematical functions, notably Fast Fourier Transform (FFT) analysis.

The Fast Fourier Transform (FFT) serves as a pivotal mathematical algorithm within modern digital oscilloscopes, enabling a critical transformation of signal analysis from the time domain to the frequency domain. While the oscilloscope's primary function is to visualize a signal's amplitude as a function of time, the integrated FFT processor decomposes this time-domain waveform into its constituent frequency components. This spectral analysis reveals otherwise obscured characteristics, such as harmonic distortion, noise content, and the presence of specific spurious signals.

For this model, a program was written for real-time data recording. Also, a program was written for Fourier analysis of the data obtained. The program interface is shown in Fig. 10. The external generator was replaced by the generator from the USB oscilloscope, the oscilloscope from the previous configurations was left for visual inspection of defects, but the same signal pattern was displayed also in the program on the computer.

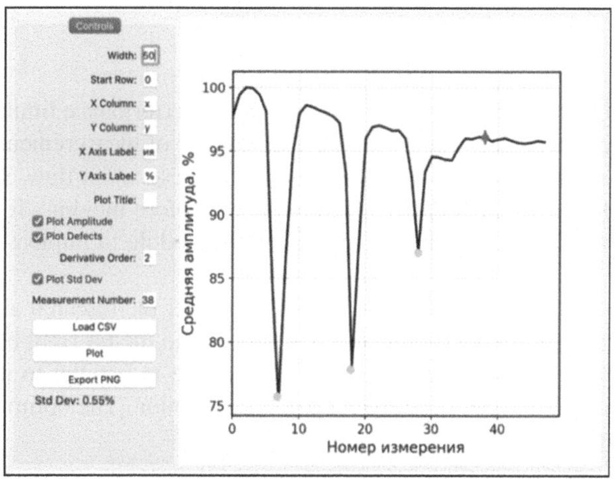

Fig. 10. Data visualization program window.

Figure 11 presents a two-dimensional graphical representation of the scan results for the object under study. The X and Y axes correspond to the sensor's positional coordinates. Signal amplitude is encoded using a color map, supplemented by a color bar for quantitative reference and isolines to facilitate precise amplitude determination in specific regions. This visualization method enables clear delineation of areas with varying signal amplitudes, allowing for visual estimation of defect dimensions and material inhomogeneities.

For data acquisition and processing, specialized software was developed using the Python programming language to interface with the module.

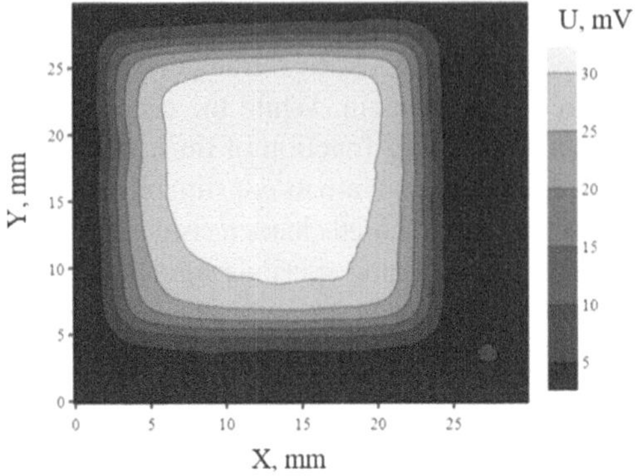

Fig. 11. Result of scanning the object of study

6 Results

Analysis of the initial algorithm revealed a significant performance limitation: the execution time increased substantially with a higher density of measurement points across the surface. This was a direct consequence of a linear execution flow, where the sensor remained idle during each data acquisition cycle before moving. To mitigate this, the algorithm was refactored using the pyThreading module to implement concurrent execution.

The scanning logic was subsequently modified. After the research area boundaries are defined, the system initiates a movement command to the far-right boundary along the X-axis. During this movement, two threads execute in parallel to manage sensor interaction, thereby overlapping communication and motion. The optimized algorithm is depicted in Fig. 12.

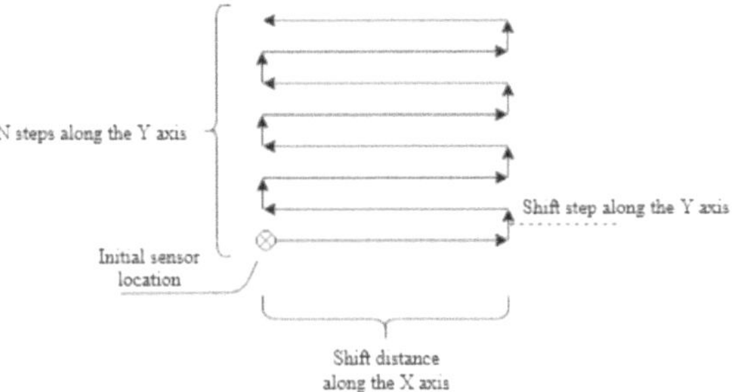

Fig. 12. Algorithm Schematic View.

Samples of the composite material "nickel aluminide - nickel", obtained by spark plasma sintering of powders of nickel aluminide grade PN85Yu15 and nickel, were studied to determine the dependence of electrical conductivity on the percentage content of components, as well as sintering conditions. Nickel aluminide grade PN85Yu15 basically contains the intermetallic compound Ni_3Al (Fig. 14).

Figure 13 shows a photograph of a sample of an alloy containing 95% of the intermetallic compound Ni_3Al and 5% of the intermetallic compound NiAl and obtained by sintering PN85Yu15 powder particles.

Fig. 13. Sample of alloy during sintering of PN85U15 powder particles.

The obtained datasets are represented as two-dimensional mappings of the voltage in the transducer's measuring circuit as a function of its position over the surface. Each data point in the resulting image corresponds to a measured voltage value, encoded according to the color scale provided. Localized defects, such as surface scratches on the specimen, are identifiable as anomalies characterized by a localized deviation in the signal amplitude from the background level.

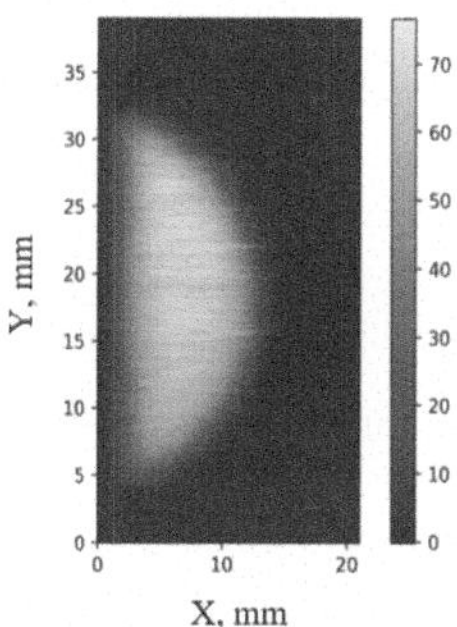

Fig. 14. Result of scanning a alloy sample of powder particles PN85U15.

The data obtained from scanning the samples are presented in Fig. 15. The specimens were produced by sintering a powder mixture with systematically varied component concentrations (PN85Yu15 powders and nickel).

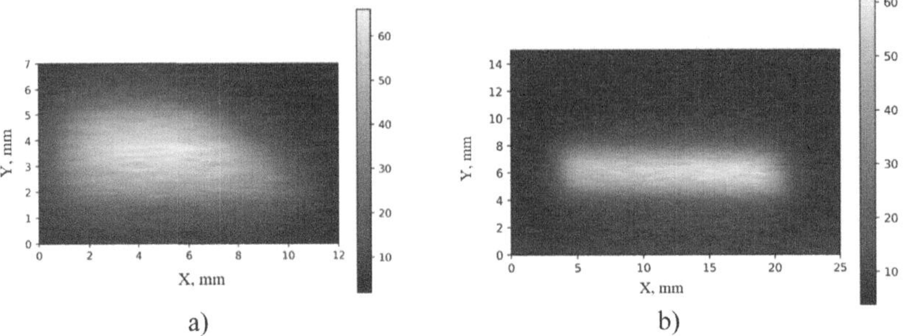

Fig. 15. Results of scanning samples with different compositions: a) PN85U15 + Ni (9:1); b) PN85U15 + Ni (8:2).

7 Conclusion

This work involved the enhancement of an eddy-current transducer and the associated software-hardware complex for its control, enabling automated scanning of test objects. The system was employed to examine nickel-aluminum alloy samples fabricated from powder intermetallics via spark plasma sintering. A relationship was established between the electrical conductivity and defects of the samples. This makes it possible to perform non-destructive testing of aluminum-nickel alloys based on measurements.

Acknowledgements. The study was implementation of the Development Program of Altai State University a part of implementation of the "Priority 2030" under the "Development of digital measuring systems for the study of metals and intermetallic structures" project.

References

1. Rocha, J.: Magnetic sensors assessment in velocity induced eddy current testing. Sens. Actuators A **228**(1), 55–61 (2015)
2. Uchanin, V.N.: Application of eddy current method for non-destructive testing of welded joints. Engine Build. Bull. **1**, 89–94 (2011)
3. Uchanin, V.N.: Multidifferential eddy current converters and their application. Techn. Diagn. Non-destr. Test. **3**, 34–41 (2006)
4. Belyankov V.Yu.: Analysis of various design options of the laid-on eddy-current transducer of the flaw detector. Bull. Sci. Siberia **13**, 41–44 (2014)
5. Jayakumar, T.: Non-destructive evaluation techniques for assessment of creep and fatigue damage in materials and components. Trans. Indian Inst. Met. **63**(2), 301–311 (2010)
6. Urakaev, F.K.: The use of combustion reactions for processing mineral raw materials: metallothermy and self-propagating high-temperature synthesis. Metall. Mater. Trans. B **47**(1), 58–66 (2016)
7. Bertolino, N.: Self-propagating high-temperature synthesis of functionally graded materials as thermal protection systems for high-temperature applications. J. Mater. Res. **18**(2), 448–455 (2003)
8. Shekari, M.: Induction-activated self-propagating, high-temperature synthesis of nickel aluminide. Adv. Powder Technol. **28**(11), 2974–2979 (2017)

9. Fan, M., Cao, B., Sunny, A.I., et al.: Pulsed eddy current thickness measurement using phase features immune to liftoff effect. Ndt. E. Int. **86**, 123–131 (2017)
10. Malikov, V.N., Ishkov, A.V.: Digital technology in eddy current scanning. Russ. Eng. Res. **43**(5), 522–528 (2023)
11. Katasonov, A.: Measuring the thickness of thin metal films using an eddy current software and hardware complex. Proc. SPIE Int. Soc. Opt. Eng. **12986**, 1298608 (2024)

Development and Research of Genetic Programming Algorithm for Structural-Parametric Identification of Dynamic Systems

Lele Zhang[1] and Nikolay B. Filimonov[1,2(✉)]

[1] Bauman University, 2nd Baumanskaya Street, 5/1, 105005 Moscow, Russia
chzhanl2@student.bmstu.ru, nbfilimonov@mail.ru

[2] Lomonosov Moscow State University, Leninskie Gory, 1/2, 119991 Moscow, Russia

Abstract. Developing a suitable mathematical model for the system under study is an important challenge in scientific and engineering research, especially when nothing is known except the inputs and outputs of the system. This problem is generally referred to as the identification problem. Different evolutionary computations based on evolutionary processes in the nature have shown high effectiveness in solving such problem. This paper is devoted to the development and study of the algorithm for identification of dynamic systems based on the processing of observed data. The identification algorithm is based on the method of genetic programming. The obtained mathematical model of the dynamic system is represented in the form of differential equations. Computer testing of the developed algorithm in Python environment using DEAP showed high efficiency. In particular, the trade-off between the accuracy and complexity of the identified mathematical models is analyzed. Also, the influence of noise on the genetic programming algorithm has been studied. The results show that the genetic programming method has good reliability when used for identification, but there is still room for improvement.

Keywords: Dynamic System · Structural-Parametric Identification · Genetic Programming · Complexity of Identified Models · Noise Effect

1 Introduction

Nowadays, mathematical modelling is one of the main tools of scientific cognition. Data and data analysis methods are trends that positively affecting the field of modeling. One of the intensively developing areas of modern control theory is the identification of systems, associated with the construction of mathematical models of systems from the observational data. Mathematical models in the form of a set of mathematical relations, adequately reflecting the basic properties of the system, are used in almost every study of real objects. For example, the first and most important stage of controlling various objects is the creation of their mathematical model, which, being the equivalent of a real object, adequately reflects its main properties [1].

V. Jordan et al. (Eds.): HPCST 2024, CCIS 2919, pp. 226–236, 2026.
https://doi.org/10.1007/978-3-032-20325-0_18

There are two types of mathematical models: theoretical and empirical. Theoretical models of systems are obtained by studying the properties of the system and the processes occurring in it. However, this is not always possible due to lack of knowledge of the mechanism of the system functioning and the theory of the processes occurring in it. In these cases, empirical models of systems are usually used, which are obtained by processing the results of observation (measurements) of external manifestations of properties and processes occurring in the system [2, 3]. Some of the most popular modern empirical methods for identifying systems are evolutionary computation methods and, in particular, the genetic programming (GP). This paper addresses the problem of developing and investigating a GP algorithm for structural-parametric identification of dynamical systems.

2 Dynamic System Identification by Genetic Programming

2.1 Problem of Structural-Parametric Identification

The problem of structural-parametric identification of dynamics systems is to extract the internal basic model from observations. And the most preferable way to describe dynamic systems is their representation in the form of differential equations. In this way, the problem of structural-parametric identification of a mathematical model of a dynamic system can be described as the problem of building mathematical model of differential equations in state variables. Here the solution of the identification problem is to find the right parts of the given equations of the identified system. To solve this problem, a set of experimental data in the form of time functions of current and initial states of the system is used, and for their control and storage it is reasonable to discretize the range of the observation process.

Suppose that a nonlinear dynamical system of n-th order described by some unknown differential equations in Cauchy form with unknown parameters is given:

$$\dot{\mathbf{x}}(t) = \mathbf{f}(\mathbf{x}(t), \boldsymbol{\theta}, t),\ \mathbf{x}(t_0),$$

where $t > t_0$ - time; $\mathbf{x} \in \mathbf{R}^n$ - vector of state of the system: $\mathbf{x} = (x_1, x_2, ..., x_n)^{\mathrm{T}}$; $\mathbf{f}(...)$: $\mathbf{R}^n \rightarrow \mathbf{R}^n$ - unknown nonlinear vector-function of the system; $\boldsymbol{\theta} \in \mathbf{R}^k$ - unknown vector of system parameters: $\boldsymbol{\theta} = (\theta_1, \theta_2, ..., \theta_k)^{\mathrm{T}}$; $\mathbf{x}(t_0)$ - initial state of the system.

As a result of some experiments on the identified system, observations of input variables (initial state $\mathbf{x}(t_0)$) and output variables (current state $\mathbf{x}(t)$) of the system at fixed moments of time $t = t_i$, $i = \overline{1, m}$ and they are contained in the matric $\mathbf{X} = (\mathbf{x}(t_1), \mathbf{x}(t_2), ..., \mathbf{x}(t_m))^{\mathrm{T}}$.

It is required to find the unknown function $\mathbf{f}(\mathbf{x}, \boldsymbol{\theta})$ (structurally and parametrically) by time data of the system's state vectors $\mathbf{X}$ that can match the input variables $\mathbf{x}(t_0)$ to the output variables $\mathbf{x}(t)$.

To assess the adequacy of the identified model to its original, i.e., the proximity of measurements $\mathbf{x}(t_i)$ experimental data $\tilde{\mathbf{x}}(t_i)$, we will consider errors in the form of:

$$\varepsilon(t_i) = \mathbf{x}(t_i) - \tilde{\mathbf{x}}(t_i),\ i = \overline{1, m},$$

and as a criterion of accuracy of system identification we introduce the mean absolute error (MAE) in the following form:

$$\text{MAE} = \frac{1}{m}\sum_{i=1}^{m} |\varepsilon(t_i)|$$

If the solution of the identified system coincides exactly with the experimental points, this norm should give values close to zero. The smaller the value of MAE, the closer the evolved model to the actual model.

As a result, the problem of system identification can be considered as a problem of finding a mathematical model of a dynamic system, i.e. a function $\mathbf{f}(\mathbf{x}, \boldsymbol{\theta})$ providing the extremum of MAE:

$$\text{MAE} \to \min .$$

2.2 Characteristic of the Identification Problem

The problem of structural-parametric identification of systems is one of the most difficult and poorly studied problems [4, 5].

In most numerical methods of identification, regression models (linear regression, polynomial regression e.g.) are used to construct dependencies approximating experimental data, the model structure (linear, polynomial e.g.) is given in advance by the research, then different numerical methods are used to obtain the best parameters for the given structure. One of the challenges of these numerical methods is that the model built with their help is essentially a black-box model, the structure of the model is unknown, and the inaccurate model structure can lead to poor results. A breakthrough result that allows us to overcome this drawback is the reduction of the identification problem to a symbolic regression problem using evolutionary algorithms for its solution [6, 7].

The task of symbolic regression is to find in symbolic form of the mathematical expression approximating the relationship between the input and corresponding output variables of the system under study. A useful property of the solution to the problem of symbolic regression is that the solution obtained, in addition to the actual computational procedure, is a symbolic mathematical expression - a formula that can be subjected to substantive analysis and then simplified or refined. At the present stage, the methods for solving the symbolic regression problem are well developed, and one of the most promising methods for finding a solution is the GP method, which is a stochastic procedure that imitates the processes of evolution and allows to represent models in analytical form. The application of this algorithm allows to obtain a model in the form of a formula suitable for further research [8]. The effectiveness of this algorithm in finding a mathematical expression in symbolic form approximating the dependence between initial and current sets of variables has been shown in a number of works [9, 10].

2.3 Structural-Parametric Identification by Genetic Programming

GP, proposed by John Koza in 1992, is a kind of evolutionary computation, belongs to the class of symbolic regression methods and is based on Darwinian natural selection and survival of the fittest [11]. The application of GP to the considered problem of structure-parameter identification of dynamic systems is based on its reduction to the problem of symbolic regression, i.e. to the simultaneous search for the optimal structure and parameters of the mathematical model of the system by enumerating various arbitrary superpositions of functions (programs) from some set. It is worth mentioning that GP automatically solves given problem without any information about the form or the structure of the solution in advance. The result of the identification will be a model of the system in the form of a set of differential equations. And the most common structure for representing individuals (potential solutions to the problem) is the tree representation of the genome form (see Fig. 1). Here, programs are represented in chromosomes in the form of syntax trees: leaves of the tree correspond to terminal (light blue ones in Fig. 1), and internal nodes correspond to functions (blue ones in Fig. 1).

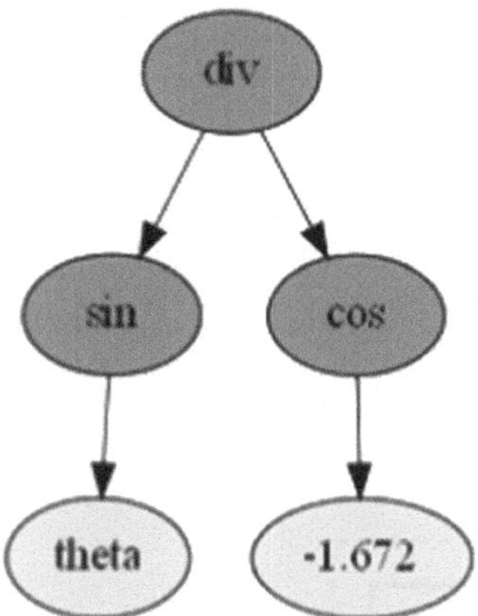

Fig. 1. Example of $\sin(\theta)/\cos(-1.672)$ in the form of tree in GP algorithm.

The general algorithm of the GP algorithm is presented in Fig. 2 [7, 12].

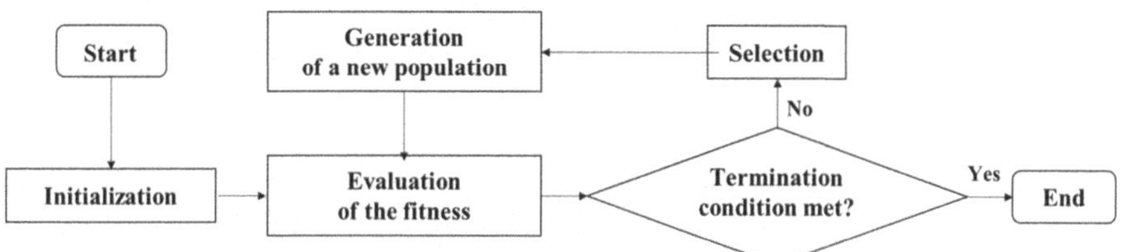

Fig. 2. General structure of GP algorithm.

It mainly consists of the flowing steps: creation of the first population of individuals by the given parameters of algorithm (each individual is one possible solution to the problem); evaluation of the adaptability of each solution by the value of the fitness function; selection of the best solutions; creation of the population of new generation through evolutionary operations (replication, crossover, mutation); checking the termination condition, making a decision to end the iteration or repeat the creation of new generation.

The general scheme of identification of dynamic systems by the GP algorithm is presented in Fig. 3. First, using the given parameters, the algorithm automatically creates the 1-st population, each individual from which is a single solution to the problem. Then evaluates the adaptability of the solution to the data, and selects the best solutions to create a new population by genetic operations (reproduction, crossing and mutation), repeats until the stopping conditions are met. The process can be seen as the problem of finding the desired program in the space of possible programs.

3 Results of Experiments

The developed identification algorithm was tested using the DEAP package for structure-parameter identification of the Lorenz dynamical system [13], described by ordinary differential equations of the form:

$$\begin{cases} \dot{x} = 10(y - x), \\ \dot{y} = (8/3)x - xz - y, \\ \dot{z} = xy - 28z. \end{cases}$$

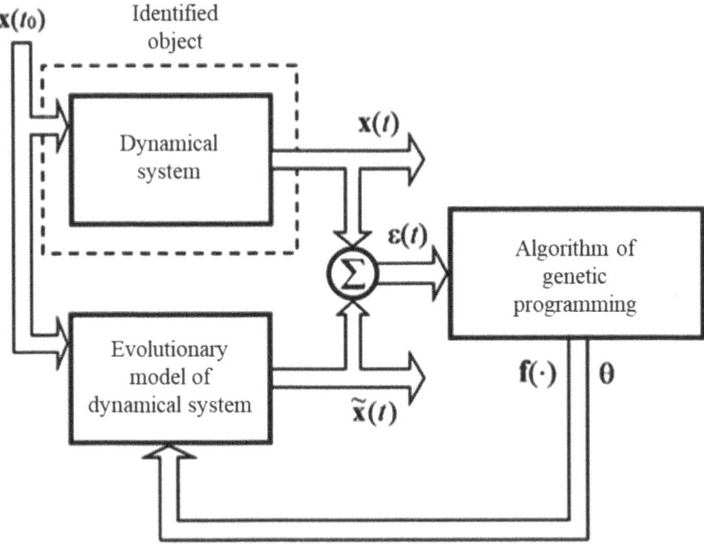

Fig. 3. General scheme of identification of dynamic systems by the GP algorithm.

We assume that the model of the dynamics of the given system is unknown and the given equations of the system are only used to obtain experimental data $\mathbf{X} = (x, y, z)^{\mathrm{T}}$ and its derivative $\dot{\mathbf{X}} = (\dot{x}, \dot{y}, \dot{z})^{\mathrm{T}}$ by integration by Runge-Kutta method in MATLAB (from $t = 0$ to $t = 20s$ with a step $\Delta t = 0.001s$). Then the experimental material is divided into two parts: the first part of observations (from $t = 0$ to $t = 10s$) is used to build a model of the system, and the second part of observations (from $t = 10s$ to $t = 20s$) is used to compare with the results obtained from the identified model.

To identify the Lorenz dynamical system, the GP algorithm was used with the main parameters given in the table (Table 1).

Table 1. Parameters for setting the GP algorithm

Generations	20
Population size	5000
Terminal set	T = {x, y, z, R}
Functional set	F = {+, −, *, /, sin, cos}
Initialisation method	halfandhalf
Crossing probability	0.7
Mutation probability	0.1
Fitness function	Mean absolute error

As a result of solving the identification problem, the empirical model of the Lorenz system took the form of the following system of equations:

$$\begin{cases} \dot{x} = 10.004(y - x), \\ \dot{y} = 28.004x - xz - y, \\ \dot{z} = xy - 2.666z, \end{cases}$$

with MAE errors: 0.0137707/0.0257665/0.00703833 respectively.

As we can see from the error result, the difference between the data calculated according to the evolved model and the real data is very minor, even the form of the models is the same with the real model, which the algorithm didn't know in advance.

To be more clearly, the verification of the dynamic of the mathematical model of the Lorenz system and its original model is presented in Fig. 4. In this picture, the solid blue line corresponds to the dynamics of the original system, and the dotted orange line corresponds to the predicted dynamics obtained from its model obtained by genetic programming. The comparison of the obtained results shows the high adequacy of the model to its original.

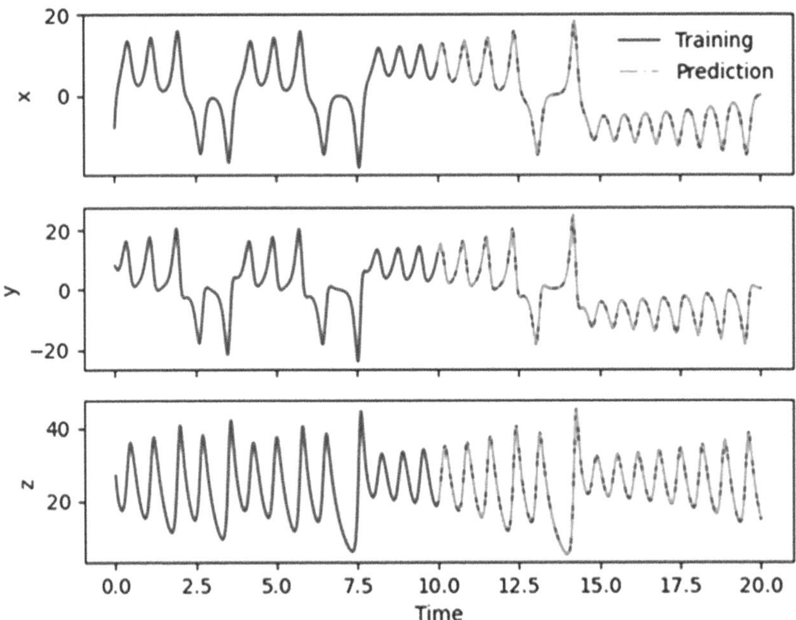

Fig. 4. Dynamics of the mathematical model of the Lorenz system and its original model.

4 Discussion

4.1 Trade-Off Between Accuracy and Complexity of Identified Models

As it is known, identification of a mathematical model of a real system is the result of a compromise between the accuracy and the complexity of its model [14, 15]. When solving the problem of identification by GP algorithm, we also obtain results characterizing the accuracy and complexity of the obtained mathematical model in the process of evolution, presented in Fig. 5. Here we can see the fairness of the well-known 'curse of dimensionality': as the number of evolutions increase, the fitness (error) decreases and model increases. In other words, the model becomes more and more accurate and, at the same time, more and more complex as generation increases. However, when revolving building models, especially for control system, some principles must be considered, like: all other things being equal, a model of minimum complexity is preferable.

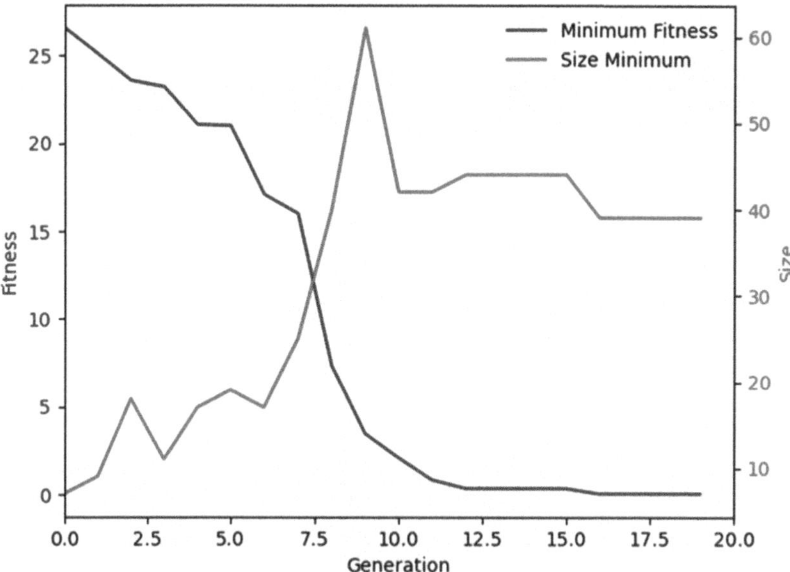

Fig. 5. Accuracy and complexity of the best identified model of the system during evolution.

To further analyze the relationship between model accuracy and complexity in the identification process, we recorded the error and size of each possible solution of x during the identification of the Lorenz system, and the pareto-front of the accuracy and complexity of the constructed model, represented by the solid blue line in Fig. 6.

In Fig. 6, in order to amplify the differences between different solutions, accuracy is represented by the negative logarithm of the error, and sparsity – the negative of the model length, this means that, the greater the accuracy value, the higher the accuracy of the model, and the larger the value of primary, the simper the model. The result shows several simple equations (e.g., with model complexity equal to -5) that predict the model of the system quite accurately. The accuracy of the model first increases rapidly at some minimum complexity, but then increases only slightly even complex equations.

How to balance the accuracy and complexity of the model is an important step in the identification of dynamic systems, especially in order to form the basic equation that contains the internal structure of the data [16]. The balance between the two must be considered when solving various problems.

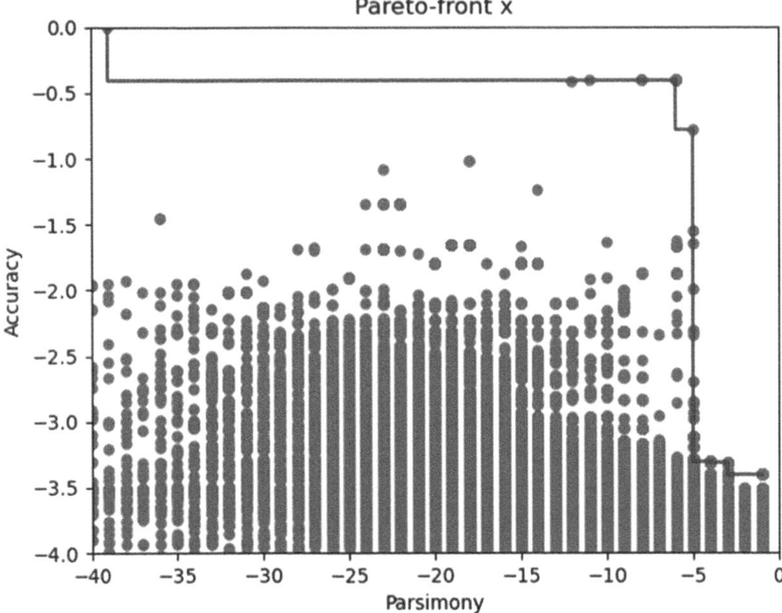

Fig. 6. Pareto-front of accuracy and complexity of the parameter x of the Lorenz system

4.2 Impact of Noise on the Algorithm

Since different noises are constantly affecting the system in the study of a real dynamical system, such as the effect of wind disturbances on aircraft flight, it is necessary to consider the impact of noise on the approach. In order to investigate the effect of noise on the algorithm, the noise data (zero mean Gaussian noise with variance σ) is added to the experimental data. The result of Pareto-front of accuracy and complexity is in Fig. 7.

Like in Fig. 6, Pareto-front in case of different noise in the data shows that despite the presence of noise in the measurement data dramatic changes in model accuracy remain at some simple models, in other words, the internal model of the system can be successfully extracted by the GP method. However, when the noise in the measurement data increases, the accuracy of the resulting model decreases significantly.

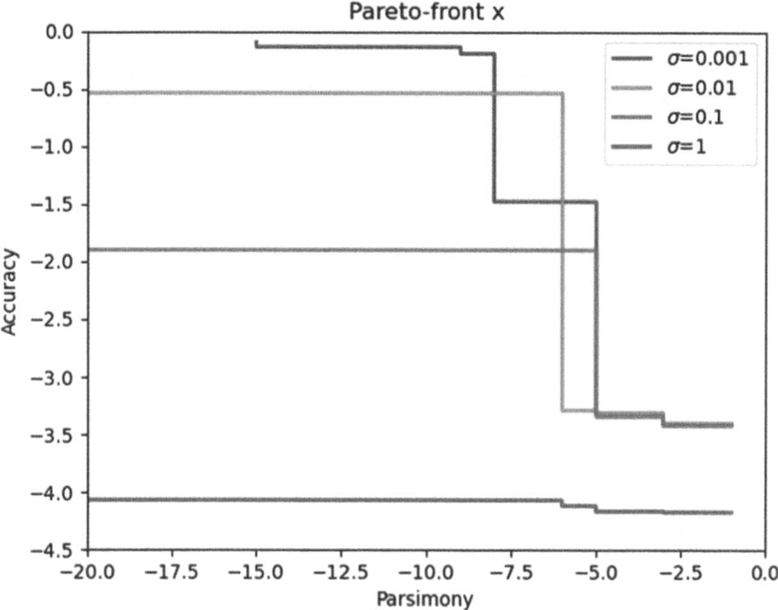

Fig. 7. Effect of noise on the algorithm in case of noise in observations.

5 Conclusion

The paper considers the problem of structural-parametric identification of dynamic systems, i.e. the construction of its empirical mathematical model, by the GP method. The algorithm of identification of structural-parametric identification of dynamic systems by GP method is developed. An example illustrating the efficiency of the developed GP algorithm for structural-parametric identification of dynamic system on the basis of experimental data processing is given.

Computer testing of the GP algorithm in the Python environment using the DEAP package has shown the existence of a compromise between the accuracy of identification and the complexity of the identified model. The pareto-front results in case of noise in observations confirmed that method is effective in extracting the underlying the internal model of the system. Another promising direction in the development of model identification methods is the balance between accuracy and complexity, since in the field of control, models are generally used for system study and control, where both accuracy and simplicity of the model are important.

References

1. Bakhtadze, N.N., Ginzburg, K.S., Borovskikh, L.P.: Identification of systems on the way to creating a general theory of identification of control objects. Control Probl. **3**, 79–83 (2015). (in Russian)
2. Deich, A.M.: Methods of identification of dynamic objects. Energiya, Moscow (1979)

3. Drozdov, A.L.: Algorithm of identification of the dynamic system characteristics by observation data. Autom. Telemech. **5**, 58–66 (2000). (in Russian)
4. Fatuev, V.A., Kargin, A.V., Ponyatskiy, V.M.: Structural-parametric identification of dynamic systems: tutorial. Izd-vo TulGU, Tula (2003). (in Russian)
5. Filimonov, A.B., Filimonov, N.B.: Structural parametric identification of linear dynamic object. Mekhatronika Avtomatizatsiya Upravlenie **23**(5), 227–235 (2022). (in Russian)
6. Eiben, A.E., Smith, J.E.: Introduction to Evolutionary Computing. Springer, Heidelberg (2015). https://doi.org/10.1007/978-3-662-44874-8
7. Poli, R., Langdon, W.B., McPhee, N.F.: A Field Guide to Genetic Programming (with contributions by J.R. Koza). GPBiB (2008)
8. Diveev, A.I., Lomakova, E.M.: Synthesis of control for group of autonomous robots with phase constraints by multi-layer network operator with priorities. RUDN J. Eng. Res. **18**(1), 125–134 (2017)
9. Karaseva, T.S., Semenkin, E.S.: Dynamic processes identification in the form of differential equations and their systems with introducing the evolutionary approaches. Herald Bauman Moscow State Tech. Univ. Ser. Instrum. Eng. **3**(144), 84–98 (2023). (in Russian)
10. Zhang, L., Filimonov, N.B.: Identification of dynamic systems based on experimental data processing using genetic programming. J. Adv. Res. Nat. Sci. **18**, 4–12 (2023). (in Russian)
11. Koza, J.R.: Genetic Programming: On the Programming of Computers by Means of Natural Selection. MIT Press, Cambridge (1992)
12. Zhang, L.: Application of genetic programming in the problem of identification of dynamic systems. In: Proceedings of the XXI All-Russian Scientific Conference of Young Scientists, Postgraduates and Students, Taganrog, Russia, pp. 399–401 (2023). (in Russian)
13. Lorenz, E.N.: Deterministic nonperiodic flow. J. Atmos. Sci. **20**, 130–141 (1963)
14. Zambrano, J., Sanchis, J., Herrero, J.M., et al.: WH-MOEA: a multi-objective evolutionary algorithm for Wiener–Hammerstein system identification. A novel approach for trade-off analysis between complexity and accuracy. IEEE Access **8**, 228655–228674 (2020)
15. Sokolov, A.V., Voloshinov, V.V.: Choice of mathematical model: balance between complexity and proximity to measurements. Int. J. Open Inf. Technol. **6**(9), 33–41 (2018). (in Russian)
16. Zhang, L.: Analysis of the trade-off between accuracy and complexity of identified models of dynamic systems. Anal. Data Process. Syst. **2**(94), 85–93 (2024). (in Russian)

Study of Power Quality and Control Strategies for Excavator Drives Using Asymmetrical Direct Frequency Converters

Ishembek Kadyrov(✉), Baktybek Turusbekov, Nurzat Karaeva, and Baktybek Uulu Azamat

Skryabin Kyrgyz National Agrarian University, Mederov Street 68, 720005 Bishkek, Kyrgyz Republic
bgtu_kg@mail.ru

Abstract. This article explores the applicability of a mathematical modeling approach, initially developed for a frequency-controlled electric drive of an ultra-high-pressure press hydraulic distributor, to the electric drives of walking excavators. The proposed model involves an induction motor (IM) powered by a direct frequency converter (DFC) built on an asymmetrical scheme with a minimal number of thyristor converters. The study focuses on evaluating the energy performance of such a configuration, especially considering the potential increase in power consumption due to distortion power caused by a reduced number of power blocks. The research begins with a dynamic analysis of the electric drive during the excavation cycle of the excavator bucket. Although the dynamic indicators were positive, they were insufficient for immediate implementation in earthmoving machinery without a comprehensive evaluation of all power consumption components—active, reactive, unbalanced, and distortion. To achieve this, phase currents of the induction motor and currents through DFC power elements were calculated and compared to those of a symmetrical system under identical conditions. The mathematical model uses a three-phase symmetrical coordinate system fixed to the stator, with rotor equations transformed into a rotating frame to maintain constant mutual inductances. This approach results in a system of linear differential equations that accurately describe electromagnetic processes and energy characteristics of the electric drive. A functional diagram of the drive with an asymmetrical DFC was developed, enabling refinement of current and voltage calculation algorithms and adjustments to the drive control system, including rotation direction control. The study includes harmonic analysis of phase currents, allowing for a detailed assessment of energy consumption. The article concludes with a formulation of the advantages and limitations of the asymmetrical configuration and offers practical recommendations for selecting appropriate control structures in electric drive systems.

Keywords: Direct Frequency Converter · Indirect Frequency Converter · Induction Motor · Thyristor Converter · Block Diagram · Spectral Analysis

V. Jordan et al. (Eds.): HPCST 2024, CCIS 2919, pp. 237–250, 2026.
https://doi.org/10.1007/978-3-032-20325-0_19

1 Introduction

The Direct Frequency Converter–Induction Motor (DFC-IM) electric drive system is widely used in highly dynamic industrial applications, such as walking excavators, to power the main mechanisms during excavation and material transportation. The shift from direct current motors (DCMs) to alternating current (AC) drives was driven by the numerous disadvantages of DCMs, including poor size-to-weight ratios, significant electromagnetic inertia of the generator, low reliability, and complex maintenance requirements. In contrast, the advantages of induction motors (IMs) are well known.

The demanding operating conditions of excavator electric drives—such as frequent overloads of the working mechanisms, full stops, high dust and gas levels, wide temperature fluctuations, and other harsh environmental factors—necessitated the development of unified monoblock thyristor converters (TCs) specifically designed for DC excavator applications. These converters were standardized to simplify maintenance and allow rapid replacement by service personnel with minimal qualifications. The term "unified" emphasizes that each monoblock has identical specifications, enabling easy substitution of a failed unit with a spare.

This paper focuses on analyzing the energy characteristics of power consumption from the grid in an electric drive system with an asymmetrical direct frequency converter (DFC), where thyristor converter monoblocks are unevenly distributed across the induction motor phases. Specifically, the studied configuration has phases A and C built according to a symmetrical scheme, while phase B lacks a converter altogether. This asymmetry introduces internal voltage distortions, compounding the already harsh operating conditions.

2 Control Algorithm for Asymmetrical LFD-Fed Electric Drive

The results obtained in the construction of a symmetrical scheme of NPHR for supplying a two-phase induction motor, in which the number of monoblocks was limited to a minimum number of modular thyristor converters, served as a basis for extending this experience to three-phase induction motors. Taking into account the practical experience and theoretical principles gained in controlling two-phase AC systems, it becomes possible to proceed with the construction of asymmetrical circuits, where the number of monoblocks remains the same as in the two-phase variant.

The asymmetrical scheme is built by connecting two phases of the motor to a reversible thyristor converter, while the third phase is connected directly to the AC network. A key advantage of this configuration in the DFC-IM system is the reduction in the number of monoblocks, while maintaining acceptable current pulsation by selecting the appropriate number of TC sets in the frequency converter. As demonstrated, the effect of the two-phase variant can be replicated by a simple reconfiguration of the symmetrical power supply schemes of the DFC-IM with a three-phase motor—resulting in an asymmetrical (incomplete-phase) configuration, as illustrated in Fig. 1.

Supplying the windings of a three-phase motor from a two-phase DFC operating in source mode generates currents $\dot{I}_{\mathrm{TC1}}$ in phase A and $\dot{I}_{\mathrm{TC2}}$ in phase C. The current flowing in phase B can be calculated using Kirchhoff's first law:

$$I_{1b} = -(I_{1a} + I_{1c}) = -\left(\dot{I}_{\mathrm{TC1}} + \dot{I}_{\mathrm{TC2}}\right), \tag{1}$$

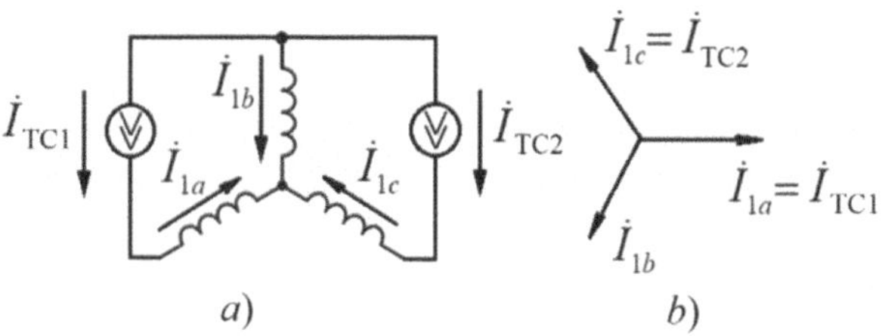

Fig. 1. The scheme of switching on two-phase current source (a) and vector diagram of motor currents (b) at connection of windings in a star

The vector diagram constructed according to Eq. (1) in Fig. 1b confirms that, from the standpoint of controllability, the symmetrical scheme with a two-phase AC supply and the asymmetrical scheme with a three-phase AC supply are functionally equivalent. However, the feasibility of applying the power scheme shown in Fig. 1a to drive the main mechanisms of the excavator's electric drive system can only be determined through a comprehensive analysis of its energy performance.

To achieve the primary objective of this study—namely, to investigate the specific characteristics of phase current formation in a three-phase induction motor powered by a frequency converter with a two-phase output—it is necessary to derive the corresponding equations for a mathematical model. This model will serve as the basis for calculating key energy performance indicators of the electric drive operating in a closed-loop automatic control system.

Assuming that under frequency–current control with an asymmetrical DFC scheme, the control law of the automatic regulator (AR) remains unchanged, and that magnetic flux linkage in the air gap should be maintained at a constant level (as in the symmetrical configuration), the formation of phase currents in the NHR should be modeled as a function of three variables: amplitude $I_m^{(1)}$, phase φ, and angular velocity ω_{0el} [2].

The equations describing the currents in phases *A* and C of the motor-drive system are derived under the following assumptions: (1) the inputs of thyristor converters TC1 and TC2 receive sinusoidally varying control voltages dependent on the three variables mentioned; (2) the windings of phases *A* and *C* are supplied from the outputs of TC1 and TC2, which operate as ideal current sources; and (3) the winding of phase *B* is connected directly to the AC mains. These equations are developed with full consideration of the inherent asymmetry in the DFC structure [3].

$$\dot{I}_{\mathrm{TC1}} = I_{1m} e^{j\omega_{0.el}t}, \quad \dot{I}_{\mathrm{TC2}} = I_{1m} e^{j(\omega_{0.el}t + 2\pi/3)}, \tag{2}$$

Using (2) we can write down the expressions of currents flowing through the AC phases:

$$\begin{aligned} \dot{I}_{1a} &= \dot{I}_{\mathrm{TC1}} = I_{1m} e^{j\omega_{0.el}t}; \\ \dot{I}_{1b} &= -\dot{I}_{\mathrm{TC1}} - \dot{I}_{\mathrm{TC2}} = I_{1m} e^{j(\omega_{0.el}t - 2\pi/3)}; \\ \dot{I}_{1c} &= \dot{I}_{\mathrm{TC2}} = I_{1m} e^{j(\omega_{0.el}t + 2\pi/3)}; \\ \dot{I}_{1a} &+ \dot{I}_{1c} + \dot{I}_{1b} = 0. \end{aligned} \tag{3}$$

where $\dot{I}_{1a}$, $\dot{I}_{1b}$, $\dot{I}_{1c}$ are vector images of currents of the IM stator phases.

The obtained Eqs. (3) make it possible to construct the vector diagram shown in Fig. 1b, which demonstrates that, when an induction motor is powered by an asymmetrical DFC equipped with thyristor converters operating in the ideal current source mode, full symmetry of the induction motor's phase current system can be achieved.

If the distorting effects of the motor's electromotive force (EMF) are neglected, then according to the description in (3), and assuming equal complex impedances in the stator phases, it is possible—even in an incomplete-phase configuration—to form a symmetrical system of fundamental harmonics in the phase currents of the induction motor.

This symmetry is achieved in a straightforward manner: by tuning the PI current regulators, which control the currents of the NPHR thyristor modules based on deviation, to their technical optimum. Notably, this tuning process can be carried out without removing the control board, which is particularly convenient for pre-operation adjustments of the monoblocks.

2.1 Control Board Function

In earlier studies involving symmetrical DFC systems, it was established that the formation of motor phase currents based on deviation control posed no significant difficulties only at low frequencies. The influence of counter-electromotive force (counter-EMF) on current shaping began to degrade the process noticeably at frequencies above 5 Hz, and this effect intensified with increasing frequency. Beyond 10 Hz, it became practically impossible to achieve acceptable current quality without compensating for the counter-EMF. In asymmetrical DFC configurations, this issue is compounded by the rigid interdependence of the two power supply channels, leading to distortion not only in the motor currents but also in the voltages. Consequently, in incomplete-phase DFC-IM systems, the need to compensate for the disruptive influence of counter-EMF becomes even more critical and is a defining feature of frequency-controlled electric drives [4, 5].

When developing control systems for asymmetrical IM power supply schemes, these characteristics must be taken into account by defining a control law for the two thyristor converters, which—as previously mentioned—are tightly coupled. Without establishing a coordinated control relationship between the converters, it is impossible to generate phase currents with the required waveform at any operating frequency.

To explore the potential for implementing disturbance compensation control, we represent the DFC as an ideal two-phase voltage source that generates a symmetrical system of motor phase currents. The specific features of DFC control are analyzed using a two-phase model of the generalized machine expressed in the α, β coordinate system. To preserve the amplitude relationships between three-phase and two-phase voltages when the stator windings are connected in a star configuration, a matching coefficient $k = 2/3$ is introduced [3]:

$$\begin{aligned} \dot{U}_{\alpha} &= 2/3 \cdot \left(\dot{U}_{\mathrm{TC1}} - 1/2\dot{U}_{\mathrm{TC2}}\right); \\ \dot{U}_{\beta} &= 1/\sqrt{3} \cdot \dot{U}_{\mathrm{TC2}}; \\ \dot{U}_{0} &= 1/3 \cdot \left(\dot{U}_{\mathrm{TC1}} + \dot{U}_{\mathrm{TC2}}\right). \end{aligned} \tag{4}$$

where $\dot{U}_{\mathrm{TC1}}$ and $\dot{U}_{\mathrm{TC2}}$ are vector images of thyristor converter voltages relative to their common point, the potential of which is assumed to be equal to zero; $\dot{U}_0$ is a vector image of zero sequence voltage.

In the two-phase generalized model, the following voltages must be generated to obtain a circular rotating field:

$$\begin{aligned} \dot{U}_\alpha &= U_{1m} \cdot e^{j\omega_{0.el\,t}}; \\ \dot{U}_\beta &= U_{1m} \cdot e^{j(\omega_{0.el\,t}-\pi/2)}. \end{aligned} \tag{5}$$

For the investigated control object, the conditions for obtaining the rotating field can be found as a result of the joint solution of the systems of Eqs. (4) and (5):

$$\begin{aligned} &\dot{U}_{\mathrm{TC2}} = \sqrt{3} \cdot U_{1m} \cdot e^{j(\omega_{0.el\,t}-\pi/2)}; \\ &\dot{U}_{\mathrm{TC1}} = 3/2\dot{U}_\alpha + 1/2\dot{U}_{\mathrm{TC2}} = \sqrt{3} \cdot U_{1m} \cdot e^{j(\omega_{0.el}-\pi/6)}; \\ &\dot{U}_0 = 1/3\left(\dot{U}_{\mathrm{TC1}} + \dot{U}_{\mathrm{TC2}}\right) = U_{1m} \cdot e^{j(\omega_{0.el\,t}-\pi/3)} \\ &\dot{U}_{ab} = \dot{U}_{\mathrm{TC1}} - \dot{U}_{\mathrm{TC2}} = \sqrt{3} \cdot U_{1m} \cdot e^{j(\omega_{0.el\,t}+\pi/6)}. \end{aligned} \tag{6}$$

Equations (6) allow us to write the system of phase voltages for the induction motor in the following form:

$$\begin{aligned} \dot{U}_{1a} &= \dot{U}_{\mathrm{T\Pi 1}} - \dot{U}_0 = U_{1m} \cdot e^{j\omega_{0.el}\,t}; \\ \dot{U}_{1b} &= \dot{U}_{\mathrm{T\Pi 2}} - \dot{U}_0 = U_{1m} \cdot e^{j(\omega_{0.el}t-2\pi/3)}; \\ \dot{U}_{1c} &= -\dot{U}_0 = U_{1m} \cdot e^{j(\omega_{0.el}t+2\pi/3)}. \end{aligned} \tag{7}$$

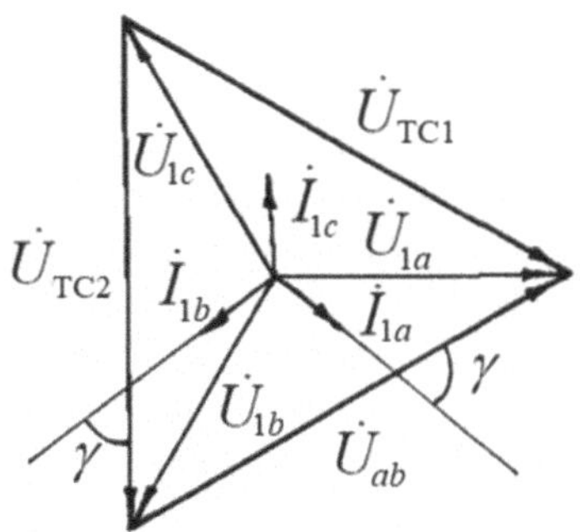

Fig. 2. Vector diagrams currents and voltages formed in the stator phases

The operational characteristics of thyristor converters TC1 and TC2 within the asymmetrical NHR scheme can be understood by constructing vector diagrams of the induction motor's currents and voltages based on Eqs. (6) and (7), as shown in Fig. 2. A detailed analysis of the vector diagrams in Fig. 2 reveals a rigid interdependence in the functioning of the two thyristor converters, which manifests in the need to assign operational priority to one of them. This priority (or subordination) is determined by the direction of the rotating magnetic field in the induction motor.

For example, if TC2 is designated as the primary converter to set the rotor's direction of rotation—similar to how a symmetrical three-phase supply defines the spatial rotation of the magnetic field by generating the appropriate voltage system—it automatically establishes the subordination of the second converter (TC1). This interrelation arises from the fact that the current from the leading voltage source (TC2) is governed by its own output voltage and the load it is connected to, while the current from the subordinate source (TC1) depends on the voltage difference between the two sources. This explains the observed equality of phase shifts (γ) between the corresponding vectors in the diagram.

An analysis of oscillograms of currents and voltages measured at the stator windings of the induction motor confirms that, while a symmetrical system of phase currents and voltages can be formed, amplitude symmetry is preserved but phase asymmetry emerges between the output currents and voltages of the thyristor converter sets for DFC.

3 Mathematical Model of Electric Drive with Asymmetrical NPD

During the adjustment of the asymmetrical NPD scheme, it was found that the main disturbance affecting the formation of phase currents in the ID is the linear component of the EMF. For instance, when the current formation priority is assigned to TC2, the rotation-induced EMF and resulting current correspond to the internal voltage of that converter. Meanwhile, the current through the subordinate converter TC1, and its associated EMF, is governed by the voltage difference between the two converter outputs.

When the direction of the rotating magnetic field is reversed, the roles of the thyristor converters are switched – TC1 becomes the master, and TC2 becomes the slave.

The functional diagram shown in Fig. 3 is derived from the functional scheme of the symmetrical system, with modifications reflecting the specific features of the asymmetrical control method. This scheme employs a three-phase microprocessor-based generator of sinusoidal signals (MPGSS), which is specifically designed to implement the frequency–current control mode of the DFC-IM electric drive. For operation within the asymmetrical system, the current reference signals are fed into the inputs of TC1 and TC2 as a two-phase system of sinusoidal voltages, shifted by 120 electrical degrees [6].

In the functional diagram (Fig. 3), the current regulators CC1 and CC2 – proportional-integral (PI) controllers responsible for shaping the IM currents through reverse thyristor converters TC1 and TC2—remain unchanged, as do the current sensors SC1 and SC2. The PI regulators are tuned to the technical optimum to ensure optimal performance of the current control loops.

The structural diagram in Fig. 3 also incorporates a control element labeled DC – a direction comparator—which accounts for the specific requirements of asymmetrical NPF control. The comparator performs two key functions: first, it generates a digital code indicating the selected direction of rotation for the induction motor, effectively assigning priority to TC1 or TC2; second, by controlling switching elements K1 and K2, it delivers the appropriate alternating currents and voltages to the EMF sensors SE1 and SE2 for the calculation of compensation signals.

3.1 Design Limitations

From a mathematical perspective, the operations performed to calculate signals proportional to the EMF generated by the rotation of the IM make the functional diagram in

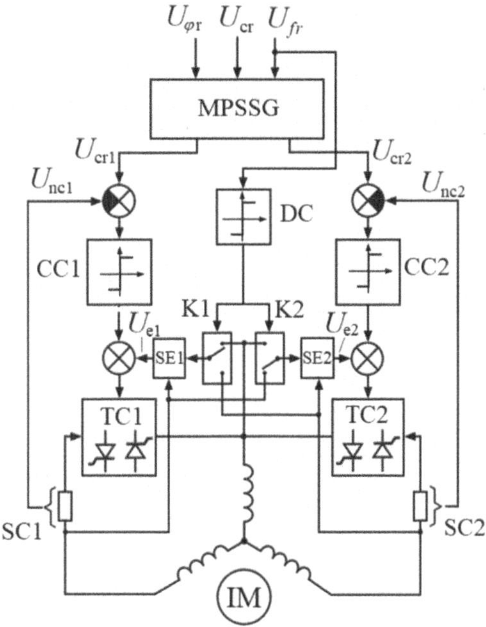

Fig. 3. Functional scheme of the combined control system, designed to form the phase currents of the IM when supplied from the asymmetrical DFC

Fig. 3 an extension of the enlarged block diagram shown in Fig. 4. This block diagram was specifically developed to support the creation of a program for calculating phase currents using a mathematical model of symmetrical systems [7].

Since switches K1 and K2 in the structural diagram (Fig. 3) determine the direction of IM rotation, the direction of the rotating magnetic field – formed by the two-phase output current system of the frequency converter – depends on their position. Furthermore, the field's rotation direction is defined by the sign of the frequency reference signal $U_{\varphi r}$, If $U_{\varphi r} > 0$, the rotation is conventionally forward; if $U_{\varphi r} < 0$, the rotation is conventionally backward.

Before evaluating energy consumption efficiency from the standpoint of power quality, it is essential to first examine the dynamic characteristics of the system, using the bucket lifting mechanism as an example – a high-tech subsystem powered from the output of the asymmetrical DFC [8]. Oscillograms of transient responses obtained from the experimental setup, shown in Fig. 5, illustrate the dynamic behavior of the electric drive under linear speed reference changes during startup, reversal, and braking operations. In Fig. 5a, the "zero load" condition corresponds to lowering an empty bucket, while Fig. 5b shows operation at rated load during the lifting of a loaded bucket.

The observed response characteristics demonstrate a satisfactory quality of speed control in the closed-loop system, which is acceptable for excavator applications. In steady-state operation, the system maintains the speed at the setpoint and accurately follows a linearly increasing speed reference. However, during transient modes, the

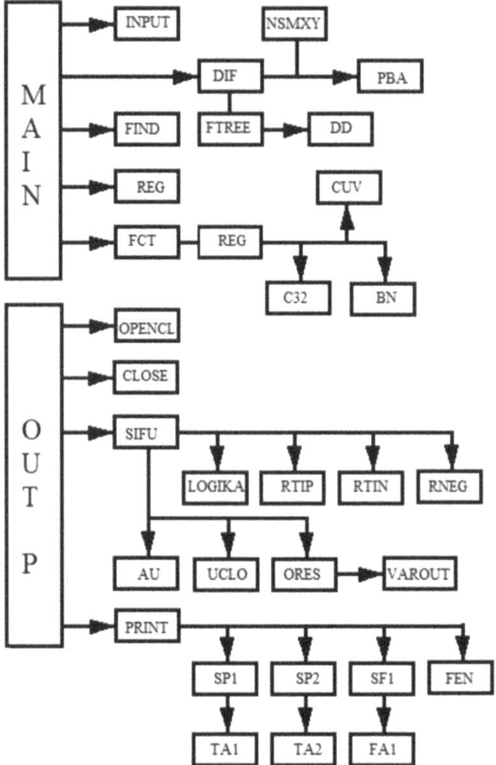

Fig. 4. Detailed block diagram for the calculation of currents in the symmetrical scheme of the DFC-IM

system exhibits noticeable pulsations in acceleration, which are reflected in fluctuations of the motor's electromagnetic torque.

These favorable dynamic indicators suggest the potential applicability of asymmetrical drive schemes in excavator systems – provided that energy consumption efficiency is confirmed through modeling. The corresponding mathematical model structure is described in [2, 7]. Without delving into the specific modifications made to the block diagram for phase current calculation in the asymmetrical DFC, the efficiency of power consumption will be assessed by comparing performance indicators from the symmetrical and asymmetrical models under identical operating conditions.

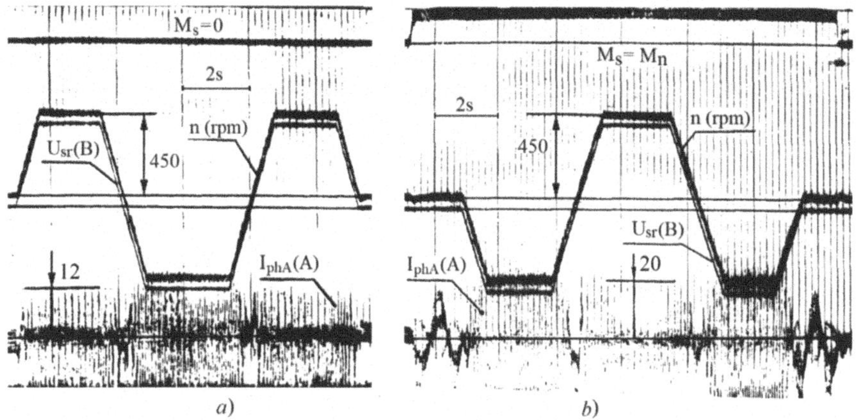

Fig. 5. Starting, reversing, braking of the frequency controlled electric drive with three-pulse asymmetrical DFC: *a*) $M_s = 0$ *b*) $M_s = M_n$

4 Estimation of Power Consumption in Symmetrical DFC

Figure 6 presents the oscillograms of calculated phase currents for symmetrical and asymmetrical DFC, shown in relative units for a single phase of the IM under different load conditions: $M_s = M_n$ (Fig. 6a, d); $M_s = -M_n$ (Fig. 6b, e); and $M_s = 0$ (Fig. 6c, f). If we disregard the waveform distortions introduced by the structural asymmetry DFC and focus solely on the smooth component of the phase current, it is found to be close to sinusoidal. This indicates that, in the case of an asymmetrical DFC, the stator currents do not significantly impair the generation of a rotating magnetic field in the induction motor [8, 9].

According to the block diagram shown in Fig. 4, current calculations for the asymmetrical DFC are performed by entering the power circuit parameters into the INPUT subroutine. The necessary matrices for phase current computation are then constructed by the *DIF*, *FTREE*, *DD*, and *RVA* subroutines. The *VAROUT* and *UCLO* subroutines are used to calculate branch currents and voltages, respectively [2].

Oscillograms of actual phase currents recorded from a laboratory test stand – configured as a physical model of the electric drive with an asymmetrical DFC (Fig. 7) – not only confirm the adequacy of the mathematical model but also demonstrate its applicability for evaluating the energy performance of the DFC-IM electric drive.

The oscillograms in Fig. 7 reveal that the conditions under which the phase current system is formed in asymmetrical configurations significantly influence current regulation quality. This manifests as amplitude asymmetry in the fundamental harmonics of the IM phase currents. Even during rated-load operation, this asymmetry reaches up to 28%, resulting in the coexistence of both forward and reverse sequences of the motor's rotating field. Consequently, the root-mean-square (RMS) value of the main harmonic components of the motor currents increases by approximately 4.7% in motoring mode, which is necessary to maintain the required electromagnetic torque [10].

The analysis of electric power consumption efficiency in the asymmetrical DFC-based drive is carried out using mathematical modeling of reactive power as an integral measure of instantaneous resistance variation over time in a nonlinear electric circuit. For this purpose, the DFC-IM system is represented as a nonlinear load powered from a sinusoidal voltage source. This representation enables the use of an expression for reactive power Q as a function of only the stator voltage and current – both of which can be calculated at any moment using a specialized computational program [2].

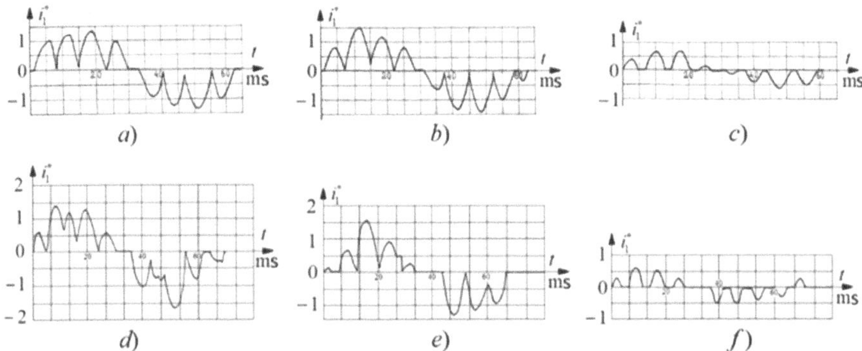

Fig. 6. Phase current of the AD: *a*), *b*) *c*) – symmetrical DFC; *d*), *e*) *f*) – asymmetrical DFC, taken for different loads: *a*), *d*) $Ms = Mn$; *b*), *e*) $Ms = -Mn$; (*c*), (*f*) $Ms = 0$

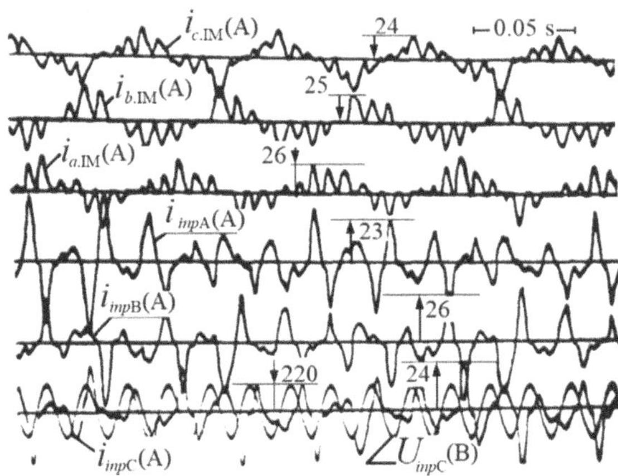

Fig. 7. Steady-state operation mode of the system with asymmetrical DFC when the stator winding is connected in a star at $Ms = -Mn$.

4.1 Results

The calculation of the integral characteristics of energy quality consumed by the electric drive with an asymmetrical DFC is performed using the *REG* subroutine. This subroutine initiates the calculation of IM phase currents by setting the initial conditions, which are specified in the *BEG* subroutine. The *BEG* subroutine inputs the parameters of the control system and the power circuit of the electric power supply. According to the current computation algorithm, this subroutine begins the simulation by bringing the system to its initial operating point.

The use of an integral data processing method enables step-by-step computation of the system's energy indicators. Harmonic analysis of the IM phase currents is carried out by the *FA1* subroutine. All values of total, reactive, and active power consumed by the DFC-IM system are computed by the *TA1* and *TA2* program modules. After processing, the results are formatted and output as final numerical indicators.

As in all valve-based electric drive control systems, the power consumption process in the electric drive with asymmetrical DFC is analyzed via software simulation, which calculates the total apparent power $S*$ relative to average values. Along with the active power $P*$ $*$, which represents actual energy conversion, the model computes the shift power $Q*$ and the distortion power T$*$, representing phase shift and waveform distortion components, respectively. Additionally, the system exhibits a significant imbalance in supply network phase loading, indicated by a high level of asymmetry power $H*$.

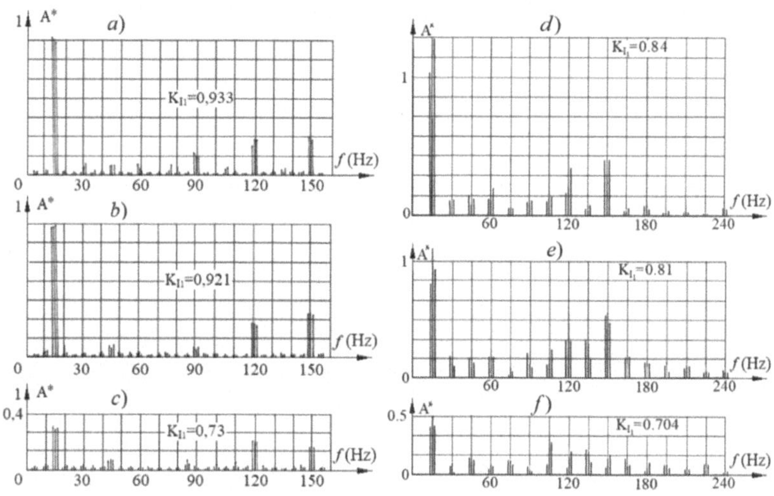

Fig. 8. Harmonic composition of IM currents when stator windings are connected in a star: *a)* *Ms* = *Mn*; *b)* *Ms* = –*Mn*; *c)* Mc = 0 all in symmetric DFC scheme; *d)* *Ms* = *Mn*; *(e)* *Ms* = –*Mn*; *f)* *Ms* = 0 all in asymmetric DFC scheme

The power consumption of the electric drive with the asymmetrical DFC is analyzed by comparing the average spectral content of the phase currents, based on the oscillograms shown in Fig. 6. This comparison is implemented using an algorithm embedded in the mathematical model of the DFC-IM system. According to this algorithm, the output currents for both symmetrical and asymmetrical configurations of the DFC-IM are computed over three periods, and the resulting spectral composition is presented as histograms in Fig. 8 [2].

4.2 Discussion

Harmonic analysis is performed for different operating modes of the electric drive—both under rated load in motoring and generating modes, and in no-load conditions. Based on the calculated current and voltage data of the IM obtained from the mathematical model, the following coefficients are derived: the power factor K_{PW}, the phase shift factor K_{PS}, the distortion factor K_{DS}, and the asymmetry factor K_{AS}. These coefficients are used to compare the power consumption efficiency of both configurations of the DFC.

The obtained numerical values of the power components, along with the coefficients that characterize the quality of energy consumption in AC electric drives, are summarized in Tables 1 and 2. A comparative analysis of power consumption efficiency for the electric drive with a symmetrical DFC is based on the histogram indicators shown in Fig. 8 (*a*, *b*, *c*) and the data in Table 1. This analysis shows that the distortion power T^* and the asymmetry power H^* make up an insignificant portion of the total power consumed. This is confirmed by the relatively high values of the distortion coefficient K_{DS} and the asymmetry coefficient K_{AS}. Across the operating range of drive loads, these coefficients remain in the range of 0.975–0.98 and 0.982–0.983 respectively at rated load, decreasing to 0.945 and 0.947 in the no-load mode.

Table 1. Averaged values of energy indices of the system with symmetrical three-pulse DFC and three-phase IM (model)

Mode	S^*	P^*	Q^*	T^*	H^*	K_{PW}	K_{PS}	K_{DS}	K_{AS}
$Ms = Mn$	2.940	1.416	2.430	0.640	0.560	0.482	0.503	0.975	0.982
$Ms = -Mn$	2.700	−0.839	2.520	0.528	0.500	−0.237	0.246	0.980	0.983
$Ms = 0$	1.120	0.079	0.997	0.350	0.360	0.070	0.079	0.945	0.947

Table 2. Energy performance with three-pulse asymmetric DFC (model)

Mode	S^*	P^*	Q^*	T^*	H^*	K_{PW}	K_{PS}	K_{DS}	K_{AS}
$Ms = Mn$	4.050	1.490	2.670	2.450	1.330	0.365	0.487	0.797	0.940
$Ms = -Mn$	3.520	−0.590	2.640	2.000	1.020	−00.168	0.219	0.804	0.954
$Ms = 0$	1.870	0.113	1.240	1.150	0.570	0.0607	0.084	0.766	0.943

Therefore, in the symmetrical system, the total power is largely determined by its active and reactive components, which explains the minor variations in the values of the power factor K_{PW} and the phase shift factor K_{PS}.

Materials in Fig. 8 (*d*, *e*, *f*) and Table 2 enable a comparative analysis of power consumption efficiency in the electric drive with an asymmetrical DFC. The histograms reflecting the spectral composition of the IM phase currents demonstrate considerable fluctuations in power consumption processes. Even in nominal operation modes, the maximum deviations in total power S^* relative to average values reach 20–25%. Fluctuations in the active power P^*, reflecting the pulsations of the electromagnetic torque of the

IM in a closed-loop control system, reach 30–33% in the same operating modes. Under no-load conditions, the active power consumption exhibits an alternating character.

The data from Fig. 8 (*d*, *e*, *f*) and Table 2 lead to the conclusion that in the asymmetrical DFC, the dominant components of total power are the phase shift power Q^* and the distortion power T^*. At the same time, the system exhibits significant asymmetry in phase loading of the supply network, as evidenced by the high level of asymmetry power H^*. In rated operation modes, the value of this component is comparable to the active power P^*, and in idle mode, it significantly exceeds it. Consequently, the system's overall power factor is low and strongly influenced by all other factors that characterize power consumption behavior [11].

5 Conclusion

Positive indicators of the dynamic performance of the system alone do not justify the use of asymmetrical schemes in high-power excavator electric drives. A comprehensive evaluation of energy consumption efficiency through simulation using a mathematical model is required.

The conditions under which phase currents are formed in asymmetrical DFC configurations significantly impact the quality of current regulation in the IM. Specifically, amplitude asymmetry appears in the fundamental harmonics of the IM phase currents, reaching up to 28% under rated load. This asymmetry leads to the coexistence of forward and reverse sequences of the motor's rotating magnetic field. As a result, it becomes necessary to increase the average effective value of the fundamental harmonics of the phase currents by approximately 4.7% to maintain the required electromagnetic torque in motoring mode.

The results shown in Fig. 8 (*d*, *e*, *f*) and Table 2 indicate that, in the asymmetrical DFC, the dominant components of total power are the shift power Q^*, the distortion power T^*, and the asymmetry power H^*. The high level of asymmetry arises from the imbalance in loading across the phases of the supply network, making H^* comparable in magnitude to the active power P^* under nominal operating conditions.

A comparative analysis of the energy efficiency of electric drives with symmetrical DFC – based on the histograms in Fig. 8 (*a*, *b*, *c*) and data in Table 1—shows that the distortion power T^* and asymmetry power H^* constitute only a minor portion of the total consumed power. This is supported by the high values of the distortion and asymmetry coefficients K_{DS} and K_{AS}, which remain in the range of 0.975–0.98 and 0.982–0.983 respectively at rated IM load.

Theoretical and experimental studies of the energy performance of the DFC–IM system demonstrate that it is advisable to implement symmetrical schemes—such as six-pulse or three-pulse circuits with three-phase asynchronous motors—when designing electric drives for the main mechanisms of medium-capacity excavators.

Based on the analysis of energy efficiency in frequency-controlled electric drives with asymmetrical NPHR, it is concluded that such drives exhibit poor energy characteristics and are only suitable for low-power applications (with rated power sub of 50 kW).

References

1. Sokolski, M. (ed.): Mining Machines and Earth-Moving Equipment: Problems of Design, Research and Maintenance. Springer, Cham (2020). https://doi.org/10.1007/978-3-030-25478-0
2. Kadyrov, I., Karaeva, N., Chyngyzbek, A.U.: Development of a mathematical model to study the energy indicators of electric drives using the DFC-IM system. Commun. Comput. Inf. Sci. **1986**, 234–247 (2024). https://doi.org/10.1007/978-3-031-51057-1_18
3. Veltman, A., Pulle, D.W.J., De Doncker, R.W.: Fundamentals of Electrical Drives, 2nd edn. Springer, Dordrecht (2016). https://doi.org/10.1007/978-3-319-29409-4
4. Bochkarev, I., Kadyrov, I.: Optimization of parameters of control device for electromechanical system of excavators. Russ. Electr. Eng. **80**, 61–65 (2009). https://doi.org/10.3103/S1068371209020011
5. Bochkarev, I., Kadyrov, I.: Microprocessor control device according to the DFC-IM system of the electric drive of an excavator. Bull. High. Educ. Inst. Electromech. **5**, 25–30 (2007). (in Russian)
6. Kadyrov, I., Karaeva, N., Andarbekov, Zh., et al.: Features of designing a variable-frequency electric drive control system with a microprocessor-based sinusoidal signal generator. Commun. Comput. Inf. Sci. **1304**, 201–217 (2021). https://doi.org/10.1007/978-3-030-66895-2_13
7. Robertson, S.D., Hebbar, K.M.: A digital model for three phase induction machines. IEEE Trans. Power Appar. Syst. **88**(11), 1624–1634 (1969)
8. Aziz, A.G.M.A., Abdelaziz, A.Y., Ali, Z.M., Diab, A.A.Z.: A comprehensive examination of vector-controlled induction motor drive techniques. Energies **16**(6), 2854 (2023). https://doi.org/10.3390/en16062854
9. Friday, I.V., Shehu, G.S., Olarinoye, G.A., et al.: Observer-based robust servomechanism control of a three-phase induction motor drive. J. Electr. Syst. Inf. Technol. **11**, 39 (2024). https://doi.org/10.1186/s43067-024-00158-w
10. Amin, B.: Induction motors in transient regimes. In: Amin, B. (ed.) Induction Motors. Power Systems. Springer, Heidelberg (2001). https://doi.org/10.1007/978-3-662-04373-8_5
11. Jahmeerbacus, I., Murdan, A.P.: Evaluation of efficiency and power quality for a two-phase asymmetrical induction motor drive. In: Proceeding of the 2023 IEEE Electrical Design of Advanced Packaging and Systems (EDAPS). IEEE (2023). https://doi.org/10.1109/EDAPS58880.2023.10468252

Author Index

V. Jordan et al. (Eds.): HPCST 2024, CCIS 2919, pp. 251–252, 2026.
https://doi.org/10.1007/978-3-032-20325-0

The manufacturer's authorised representative in the EU is Springer Nature Customer Service Centre GmbH, Europaplatz 3, 69115 Heidelberg, Germany. If you have any concerns regarding our products, please contact ProductSafety@springernature.com

Printed and bound by CPI Group (UK) Ltd, Croydon, CR0 4YY

07/07/2026

02160913-0006